FROM WHENCE COMETH
INTELLIGENCE

SYDNEY L. HERRERA

FROM
WHENCE COMETH
INTELLIGENCE

SYDNEY L. HERRERA

CITI OF
BOOKS

CITIOFBOOKS, INC.
3736 Eubank NE Suite A1
Albuquerque, NM 87111-3579
www.citiofbooks.com
Hotline: 1 (877) 389-2759
Fax: 1 (505) 930-7244

Ordering Information:

Quantity sales. Special discounts are available on quantity purchases by corporations, associations, and others. For details, contact the publisher at the address above.

Printed in the United States of America.

ISBN-13:   Softcover   979-8-89391-378-1

                          eBook   979-8-89391-379-8

Library of Congress Control Number: 2024921203

TABLE OF CONTENTS

ABOUT THE AUTHOR

Born in Siparia, Trinidad and Tobago, one of fourteen siblings, first to attend high school, extremely trying upbringing including being left at various individuals because of hardships, outstanding student even though getting food or clothes was challenging throughout his school age years.

Had a "near-death" experience in early teens in which he heard a voice say he had to go back because "no one knows where he is and his work is not done yet". He believes his education and life's path is related to, and strongly influenced by this experience.

Graduated high school then worked for several years in the oil industry, left Trinidad in 1984 to learn computer programming in London, England. Emigrated to US in 1986, and has since worked as a software engineer.

As a software engineer, worked on many systems in both private and governmental agencies, on mainframe and distributed environments, one of the few who are expert on mainframe, middleware and distributed computer architectures and software applications. Specialized in modernizing legacy applications transforming legacy applications and migrating from relational databases to big data warehouses both on-prem and cloud based. Winner 2021 Innovation Award from MongoDB for an AI software application that he designed and built.

Inspired Books:

Spiritual Renewal – renewing our pact
Understand, Improve – Self and Society
Humanity's Inflection Point
From Whence Cometh Intelligence

Preface

"Every intelligent action occurs as a result of logic"

"All changes witnessed in evolution are directed by logic and accompanied by the imbedding of logic in living organisms to support the successful implementation of those changes."

"There must be an Internal Processing System (IProSys) that like DNA is intrinsic to every living organism and may also be part of every cell, it may be what constitutes "life" itself!"

This book is the corollary to and fills in the gap that was created and has not been addressed since the publication of "On the Origin of Species by Means of Natural Selection" by Sir Charles Darwin in 1859, the gap is the evident logical progression in the DNA of animals along with the seeming inherent knowledge that accompanied the physical enhancements that enabled diverse species to evolve successfully. This unaddressed gap in knowledge is what results in naturalist and even our best and brightest no longer referring to random happenstance being the driving force in evolution but have now coalesced around "Mother Nature" or simply "Nature" as the logical driving force as we have become more knowledgeable about the advanced and sophisticated design concepts that are evident.

The subject of this book is intrinsic intelligence, not that which is taught or acquired by experience of observation but that which is common referred to as "instinct".

The dawn of our Intelligence age must raise the curtain on the fact that without intelligence, there is only chaos, it must also bring the curtain down on any notion of random being a part of our existence or ecosystem. Since we have started building Artificial Intelligence which enables us to empower our creations with the ability to perform tasks autonomously or semi-autonomously, it is now apparent that without intelligence every activity and process would be analogous to starting a car and just letting it run or putting it in gear and letting it go on down the road, in the first case it just sits there and runs until it's out of fuel, in the second it goes on in a set course until it runs out of fuel or collides with something, with intelligence however it can perform useful task unattended, it can become a robotaxi that can be summoned to transport occupants autonomously, similarly nothing in Nature functions without intelligence; from the unicellular organisms to the most complex, there are no random, chaotic actions, rather what we can observe everywhere at every level, are activities directed and guided by intelligence.

We can now recognize that intelligence is what enables the inanimate to be transformed into something that performs any action in an organized manner, as without a decision-making process, actions would be chaotic. When energy is applied to anything, it would move in straight lines until it encountered something to change its' direction, essentially lunging from one crash into another, but as we know, nothing in Nature behaves this way.

We trust our sciences because they based on empirical measurements of observed phenomena, unless we currently do not have the means to make measurements that can be independently verified as is the case currently with logic embedded in living organism, until we are able to measure, we must commit to accepting consistency based on data from extended observation. This is the case with recognizing the "programming" evident in Intrinsic Intelligence. It is based on extrapolating existing rules and our understanding of the infrastructure that is required to create, store and execute logic, along with our observation of the consistency in the execution of logic that is evident in all subjects studied.

There is a literal binary choice, like choosing either zero or one, we either must accept that all plants and animals have the ability to self-engineer every aspect of their physical form, that is to say they must be able to manipulate their own DNA so that the changes they "make" are passed down to their offspring as well as updating any changes in their instincts that may apply to those specific physical changes or, those changes that occur the DNA and instincts where relevant, are totally beyond the control of that particular plant or animal.

This book takes the view that the changes that occur to DNA and instincts are totally autonomous, and beyond the control of any individual plant or animal.

Intelligence is the Difference

"Without intelligence, actions if they occur, are random and chaotic"

Our civilization has and continue to advance as a result of the work of dedicated men and women and our ability to collaborate and document the discoveries of dedicated individuals everywhere, it is important to note however that our knowledge is the product of discoveries and not the creation of any one or group of individuals even though we may celebrate their hard work and dedication by having their names associated with any resulting process or innovation. We may have invented languages and the ability to transcribe our thoughts and activities but the discoveries that gave rise to our great civilization are all the result of the truths about the forces and fundamental properties of resources that existed for time immemorial before their "secrets" were apparent to us.

We are fortunate to have a written tradition that enables every generation to retest established rules and modify them, if necessary, as well as expand and build upon them, thus expanding our understanding of the forces, material properties as well as any rules that govern the behavior of fauna and flora throughout the world. Special mention has to be made of the excellent work done that resulted in the documentary video "In the Mind of Plants" which is on YouTube because it breaks ground in the exploration and recognition that plants like animals have memory, it also delves into plant intelligence.

Scientific exploration and the establishment of accepted principles are normally based on empirical data that can be independently verified, there are however fields where there is absolutely no measurable data but accepted rules are based on repeatedly observed phenomena, examples are the widely accepted prognostications of psychiatry and psychology. With the exception of the two aforementioned disciplines, science requires that for discoveries to be accepted, there must be data that can be independently verified and tested before being accepted, there are of course exceptions that must be made when there are clearly observable phenomena but the scientific methods to produce any measurements are not currently available.

Our sciences advance as a result us observing the **tangible** which evokes our curiosity and by asking "Why?", the revelations of Physics and Chemistry are brought about, by those who ask "Why does this occur?", "Why does it occur in this way?", the answers to those questions enable us to truly understand "What" lies behind that which we observe and enables us to understand and predict "How" things will unfold. The answers to the "Why?" often leads to the **intangible** part, which is the driving force behind the behavior of what we observe. Seeking the answers to "Why" even enables us to manipulate resources based on the understanding of "How", so that we can harness electricity, use chemical reactions to create all the products that power our civilization.

Our current theory of evolution address only the observation of the tangible, what has been missing, is the understanding of the intangible, that which is the driving force for what we observe, it is the "this occurred because of xyz". What we have done in this branch of science is instead for seeking that intangible based on the logical way that everything unfolds as we do in other branches of science is to relegate the logic we observe to a mythological "Nature did it", if we did the same in Chemistry or Physics, we would not have had all the time and dedication of so many men and women that resulted the pursuit of these intangible forces that has led to all the discoveries that are enshrined in our Laws of Physics and Chemistry, instead, everything we observed would have been relegated to "Nature did it".

We have entered a very transformative period in our civilization, the period will be recognized in our history as the period where we realized that the behavior of every living thing is driven by logic and just as there are forces everywhere in nature, from the subatomic to electricity and plate tectonic, and these forces are governed by rules that we have investigated and codified, the existence of all living organisms are governed by rules of logic. So, it is impertinent to assert the existence of any living creature to be the product of random happenstance as not only is its evolution the product of logic, but logic is embedded in every living organism, and all their actions and reactions are directed by logic.

Our recognition of that logic and intelligence governs the living poses many challenges that may seem insurmountable but, that's where all the challenges of our sciences start, it requires individuals of indomitable spirit to persevere and make the breakthroughs that will advance our civilization to unprecedented heights.

Our use of logic started with the desire to automate repetitive tasks and facilitated by our advances in computer technologies to enable the development of Artificial Intelligence (AI) to enable some of our machines to operate autonomously or semi-autonomously. The recognition that intelligence is what is required to transform machines from pieces of equipment lying around waiting for an operator, to robots that can perform all manner of tasks without requiring any input from a human operator is the key to unlocking one of the deepest mysteries of our world.

It is our understanding of what goes into the making of intelligence, from the infrastructure to the software components that will be transformational in our understanding of many of the mysteries that have had science reaching into the untenable hypotheses that "Nature did that" to explain all the logical elements that are evident throughout our ecosystem. It is also evident that this is a natural progression, even with some of the disruptions that will undoubtedly occur, this is also the natural progression of science, from observation and recognition to achieving a true understanding of the underlying processes.

Our scientific advancement has enabled us to begin emulating albeit in a very limited way, the unknown forces of the nature and possibly of the very universe. This is our ability to create intelligence which we refer to as Artificial Intelligence (AI) which we use to enable our machines to operate and even learn from events related to their functioning.

Our forays into AI now coincidentally enables us to discern the very building blocks of intelligence, we now know that there must be memory in which to store information and advanced information is normally built and then must be embedded in the memory of the subject, or some mechanism must be provided to enable access to memory that is stored externally to the individual machine. Our AI is possible due to the advances in the size and speed of memory that we have achieved through years of computer related research.

Almost all our scientific discoveries are related to the discovery of the Periodic Table of Elements which the discoverer Dmitri Mendeleev stated was revealed to him in a dream and which he initially published in 1869. This is directly related to all the materials used in our modern-day research, as well as almost every facet of human existence; sub atomic forces and even DNA could not be discovered without the research tools created based on our knowledge of the properties of metals etc., even the fuels we use to provide energy is based on this revelation of this bedrock of science.

We can now recognize that we may have missed one of the greatest synchronicities of human existence by not having the knowledge, to see the relevance of a publication ten years before the revelation of the Periodic Table. The publication by Charles Darwin of "On the Origin of Species by Means of Natural Selection", the primary work that resulted in the Theory of Evolution, did not have our understanding of intelligence as an intangible attribute that is responsible for enabling the inert to perform meaning tasks. It is only now in the twenty first century and with our work in the field of AI that we can now recognize that which was missed during that period of human enlightenment.

It does not take a lot of reflection to recognize that intelligence provides the directives for the actions of all living things, without intelligence there would not be the patterns that we observe in all plants and animals that enable us to create categories of organisms related to their physical characteristics and behavioral pattern, we know that DNA is responsible for the physical characteristics, intelligence is responsible for all behaviors including growth patterns.

Without intelligence we would not have the ability to differentiate and so would not reliably react to changes in the environment around us, without intelligence we would not know the difference between up and down, light and dark, hot and cold, and neither would plants, all the behaviors of plants and animals are the product of intelligence, it is also very clear that there are different types of intelligence, the type that is the product of learning whether by observation, practice, being taught by others or by any other means and that knowledge that is inherent in plants and animals from their inception and is common to every member of each individual specie.

Our recognition that intelligence is present in all organisms has always been tacitly accepted but what has been missing is the realization that intelligence is a separate and distinct characteristic or every living thing, just as distinctive as hair or skin or bones, and an individual organism could not exist as we know it without the built in intelligence. The very significant difference for the entire theory of random evolution however is challenged by the lack of evidence of true random, chaotic relics in nature, but more by the direct conflicting evidence of the development of specialized intelligence which complements every living organism and is somehow permanently embedded into their memory.

All the data in the world residing in a storage system that is offline is still information, same with any of our robots such as a car or drone that is turned off, so data by itself is not intelligence, a processing "engine" is required to be able to use all that information to make logical decisions before it can be viewed as an intelligent system, so without the processing capability making use of accessible data, no intelligent system would ever exist. So, every living thing has to have some form of processing capability and a store of reference information to make intelligent decisions that can translate to intelligent actions.

Future research may show that the "processing system", which will be referred to as the Internal Processing System (IProSys), is a distributed, networked "computing" system, which has components in every cell, much like DNA itself, it may be directly associated with DNA as its' functions are closely coupled with DNA and the directs the activities at a cellular level, with concurrent communication across the entire organism.

The IProSys may indeed be included in the DNA or in another system that is intricately associated with DNA, but as of now we recognize that DNA is essentially a "blueprint" or as a "reference database", it does not have the "processing capability" to make decisions and then direct other cells to perform those directives. It is as if by some unknown means we have a plan for a city (not just replication since different actions have to occur at different times, with significant coordination!), some entity has to direct and coordinate all activities to ensure the successful completion of the project.

Just as one of if we were to create an intricate and detailed design, it does not in itself result in the completion of an end product until work directed by intelligence completes all the steps to bring to fruition the product detailed in the plan.

Analyzing the very beginning of life for most complex animals demonstrates the need for this processing capability and this occurs even before a brain is present! After fertilization the IProSys indicates that cell division should start and likely coordinates the production and delivering of "materials" including DNA to every new cell produced, however the IProSys has to monitor the cell splitting/new cell creation process because at specific times it has to indicate/direct/control the creation of network of nerves, a system for the transportation of nutrients, likely the creation of an endoskeleton and or an exoskeleton, a digestive system etc. The database or blueprint DNA must be constantly replicated and referenced but the details of the IProSys while it is essential, are currently not understood.

DNA is the "blueprint", the architectural directions that must be followed for every organism physical characteristic, DNA however does not DO anything on its' own, even the inclusion of DNA within each cell and the very process of creating new cells based on the DNA blueprint has to be accomplished by evidently intelligent processes that "use" the data stored in DNA. Even the chemicals that may be used in signaling changes have to be created under the direction of the IProSys, there has to be an intelligent "agent" to "read" that blueprint and direct the execution of the process of creating new cells based on those directives, so even in the process of cell division and growing new cells, there is the intelligent conforming to the information contained in our DNA.

Every process that is occurring in every living organism has to be monitored and maintained, this work is done by the IProSys.

Even some aspects of physical characteristics are the product of this inherent intelligence, grasses for example as opposed to other plants have fibrous roots, so based on the information in the DNA "database", new roots are created with the characteristic attributes rather than a main root, the information for these characteristics are stored in the DNA but the IProSys is responsible for "reading" the blueprint and directing activities based on this data.

The actual formation of an embryo and the generation of cells that determine the physical form of the animal is again the product of intelligent processes i.e. an intelligent "agent" has to be active from the very first cell division to determine based on information in the DNA "database", 'are we generating a heart, lung, bones, brain, appendages, skin, blood vessels etc.', even the sequencing and timing are all the product of this intelligent agent which we must note is providing this direction and coordination even prior to the formation of the brain.

Of course, all autonomous functions occur with evidence of intelligent decisions being constantly made for the benefit of the organism, in animals this includes recuperation and regeneration during periods of rest. That these activities occur consistently across any specie is proof that this "programming" is indelibly written as consistently as DNA even as we are currently unable to determine its' location.

We must recognize that there are sensors that can detect cell association, as we can observe in the operation of stem cells that can "recognize" proximity and create cells accordingly, even chemical signals require intelligent processing to "decide" on appropriate actions.

We must now recognize instinct is in fact Intrinsic Intelligence (II) which is best defined as "**evident knowledge that is not obtained from external sources",** and is an element present in all living things, this information has some of the same elements that are essential to the retention of AI, even though II is eons more advanced and resides in living organisms versus AI in our machines. II is NOT based on any information acquired from external sources, most human intelligence is NOT II, II however is fundamental to life itself and can be observed in all humans in the autonomic processes, the decision-making processes that occur within our immune and digestive systems, growth and healing, the operation of stem cells etc. are all dependent on the II that is embedded within each of us.

We recognize II everywhere in Nature, it is that which we refer to as "instinct", it is that knowledge which has no discernible source but which is so pronounced that it directs the actions of all animals, to reiterate; this is NOT the behaviors learnt from association or observing others but examples would be web spinning by spiders, intricate nest building by birds etc.

All the changes witnessed in evolution are directed by logic and accompanied by the imbedding of logic in living organisms to support the successful implementation of those changes.

We can now recognize that the development of the information required for II must have taken quite considerable periods, this would line up very closely with what we now know as the evolutionary times, we also know that the processes involved in intelligence do not have to be "consciously" known to the "beneficiary" of that intelligence, for example a car or drone does not have to "know" which processors and what specific code is being executed by any of its subsystems like radar, lidar, GPS, gyroscopes etc., in "intelligent machines" these systems all work towards the proper functioning of the machine without direct action of the machine, although their activities more or less dictate the ability of any machine to successfully complete any assigned task.

There are of course significant differences between AI and II, the most significant being of course that AI is implemented in non-living machines, and although we may hear about machines acting lifelike, that machines cannot reproduce will always relegate them to being dependent on their "creators" for replacement parts and without perfect coding, even software updates, II by comparison having been "perfected" over eons and implemented in living organisms in worst case scenarios lead to extinction but more likely results in the "evolution" to adapt to changes so as to be able to take advantage of even unforeseen changes in the environment without direct action of the creator especially if the creator is aware of what may have been unforeseen to the all that exist within the ecosystem.

We can now trace the evolutionary paths of most species, and although it is readily apparent that there are intelligent forces at work based on the direction and results evident everywhere in the evolutionary process, there is a complete dearth of information on the logical system or systems that are responsible for the results we observe. We now attribute all logical processes to Nature which is one of the most accepted oxymorons in that random Nature is responsible for the logic and order that we observe everywhere within it!

Our advancement to be creators of logic that we can implant into our machines, now makes us aware of what goes into the creation of logic as well as an awareness that logic governs everything around us, from the stable forces that determine the characteristics of atomic and subatomic particles and the activities of every living cell to the most complex symbiotic relationships between plants and animals or even those between earthly and terrestrial entities. As a result of these consistent patterns, we are able to predict activities based on observable patterns which led to my hypothesis: "The existence of governing principles (understood or not) that enables the accurate prediction of results or outcomes nullifies any notion of random in the related activity".

Our development of logic which simply put is the development of decision making based on evaluating a series of conditions with the intention of achieving some desirable outcome, enables us to now recognize this decision-making process pervades throughout Nature, so we can state emphatically that nothing is random in Nature.

If we were able to create machines that could reproduce themselves, with machine learning within all their subsystems, of course with updated processors and memory subsystems also, our machines would essentially have free will they could operate truly free of our interaction, we would then have to instruct them of a code of conduct that was designed to in effect to protect the ecosystem that is essential for the balanced proliferation of not one but all species and the very ecosystem. We would also have to make critical decisions and our intervention would occur if and when specific species having free will, by their deliberate actions, through ignorance, greed etc. they consciously and consistently strive to degrade the ecosystem to the detriment of all others.

It must also be apparent that it would be beneficial to all if at least one specie could be elevated to be stewards, to this specie could be imparted not only principles to be followed, although the ecosystem is itself could have the ability to be self-correcting over time. The appointed stewards would need not only specie specific II but also over time, many of the secrets of the system would be revealed so that they can be more effective in their elevated role. In a fair and balanced system, elevated capabilities must also be paired with repercussions to curb any abuses.

Elevation of any group would be accomplished through education of the laws and methods used in the actual creation of the entire ecosystem.

Our current definition of evolution simply ignores the intelligence we can observe in our ecosystem and the II that we call instinct that is inherent in every organism, as it stands the definition seeks to address the physical transformation that we can observe which was fine until we now recognize since we have started installing Artificial Intelligence (AI) into our machines, that without embedded intelligence, every organism would simply have to learn to navigate within our ecosystem, for all the "lower" animals who do have some sort of parental tutelage, their actions would be completely random and chaotic.

So our definition of evolution completely ignores the fact that analogous to our computer systems, there is a hardware component and even more important when it comes to intelligence, a software component, it is the software which makes all our machines "come alive", it is the part that makes decisions, so in all animals, not only is there that physical component which we can observe but all their actions are in fact controlled by this "software" component, it is the part that does the "reasoning" and results in the logical activities of every living organism.

And so, at this point in our civilization we have to have to address this question: Can the logic that guides the processes we observe in evolution and which is also embedded as unique and distinct II within every living organism be the product of random happenstance?

We know that intelligence is the built from posing and answering a series of questions which can be demonstrated as a series of "if condition exist then perform some action A else perform some other action", which we can call B, both action A and B could be themselves or contain other checking other conditions and performing other actions as demonstrated in the following diagram, which shows some initial processing e.g. is read available initial data (Process A) then make a decision based on current state (Decision 1) this then leads to choosing one of two paths (Process B or Process C), each path can then have other processes and decisions.

In living organisms of course, the "questions" asked may be more nuanced as in plants after checking light sensors which transmit information of light intensity and direction, it may be "Which direction is the light coming from" with the resulting process being "Initiate cell growth at specific points to turn leaves towards light", a process that we observe as phototropism but which without logic is simply not possible.

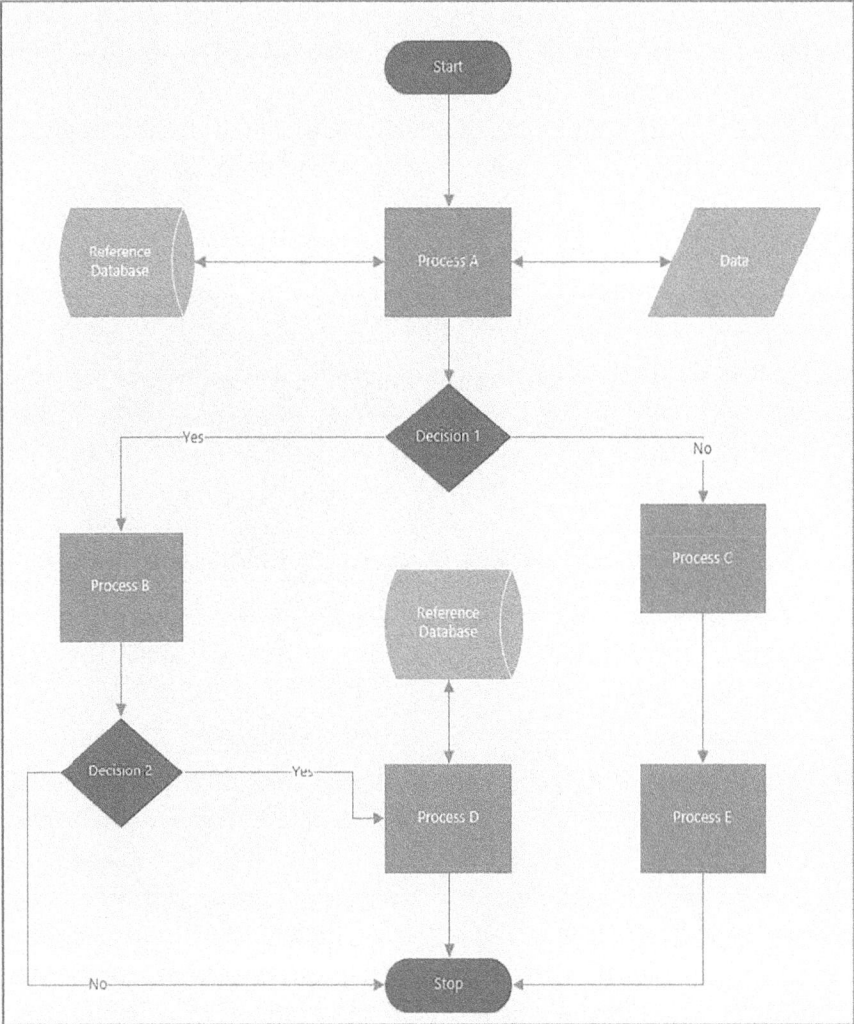

If *condition X exist*
Then *perform action A*
Else *perform action B*

Action A and B could indeed have similar constructs interrogating other conditions with the same IF/THEN/ELSE structure, condition X could itself be the result of evaluating other conditions. We see this application of logic everywhere in Nature, from plants that interrogate sensors to determine the source of light or water with resulting action of cell growth to target sources of nutrition to producing color or scents to indicate to animals that there is food and thus have the animals be transportation mechanisms for their seeds to ensure dispersal of their genetic material.

We can observe logic specific to sets of organisms within the same species, as in bee or ant cultures. It is evident the various types of ants do not get any form of training to perform their functions, and thus we must assume that the "programming" of the logic is somehow hardwired into the queen, drones and even those that may generate wings to be able to establish new colonies under certain conditions.

The intelligence that we will focus on is that which is NOT learned but exist within every organism. Within ourselves, this type of intelligence is manifested every time we are at rest, the healing processes all are directed and/or controlled by this advanced sophisticated logic making use of chemical and other sensors to work unconsciously for our benefit, similar actions are always being caried out by our immune system, using logic and intelligence embedded within each of us. There may also be logic and or sensors that respond to elements that are totally outside our bodies like magnetic or even terrestrial bodies.

It is apparent that every living organism has some embedded intelligence, from the unicellular to the most complex, each has specifically tailored as well as some elements of common logic.

Our experience with developing AI gives us insight into some of the prerequisites and infrastructure required to store information that comprise the logic but unfortunately this does not translate or provide any inkling into the data stored in any living structure or organism. Our information systems currently are based on digital data, information stored in living cells however are a complete mystery to us at this time, we are completely in the dark as to where, how or even what is stored and as such the mechanisms for retrieving this stored information can only be the subject of theoretical exercises. We do know however, just as we can observe the sun rise and set, that information is stored and retrieved and comprises a highly sophisticated system that may extend outside the physical extremities of organisms.

When we observe large swarms of flying insects, flocks of bird or schools of fish that travel in very close proximity with almost instantaneous changes of direction at relatively high speed without catastrophic collisions, it is apparent that there are sensors and information being processed that involves simultaneous actions by many if not all members of the group, logically this infers that somehow data is being transmitted instantaneously and possibly autonomously to multiple members of the group.

The logic which we are now experts of, is the result of "asking" a series of questions and making the best decision based on currently available information, making the best decision has improved from simple computing or addition of numbers to having complex databases that can now include information based on "experiences" that were encountered after the initial data is "loaded", this phenomena which is referred to as machine learning, can result in the autonomous accumulation of "knowledge" so that the decision making process becomes more finely tuned and the resulting decision making seems "wiser" based on this increasing store of "intelligence" over time.

One of the concepts that many may have difficulty accepting, is that of true "free will" or fully autonomous systems, this is likely due to our recognition of our own failings or shortcomings. It is inconceivable to many of us that we would truly program a robot and not give ourself the ability to step in at any moment and take control of that robot or even be able to turn the power off due to our recognizing that our robot may malfunction, this is the fear that causes major pushback when we learn that scientist are experimenting with living organisms, because unlike our mechanical inventions, we do not have an "off" switch for living organisms once they move out of our physical control and we do not have the ability to predict what mutations we may initiate.

It is very clear that almost everything we encounter in our lifetime can be explained by logical processes, all our actions as well as the actions of everything around is the result of logical processes, the actions of every plant or animal are the result of intelligent logic. We can all validate that every action we encounter has a logical basis by simply by posing the question of "Why did this occur?", as simple as this sounds, it leads us to recognizing that every action is the product of logical decisions. Even natural phenomena like weather events, earthquakes, tsunamis, tornadoes can be explained logically as whereas this book does not venture into that realm, staying with living organisms, the IProSys and II, we can recognize that every action taken by every living thing is the result of decision making, when an animal wakes up, when it goes to bed, every action they take during their entire lives are the result of conscious decisions, which in most cases we can determine.

We can recognize that different forms of our overly simplified *"IF this condition exist DO something ELSE do some other thing"* is fundamental to every living thing, when however, we can recognize that the scope of the resulting activities can be clearly defined across any specie, that II which we call instinct. We must recognize that not only is the process the same but, the information used to make the decision is the same, since individual animals are not in contact the information must be stored in their memories, and thus making decisions based on the same information results in the same "instinctive" reactions, which is what we observe.

Our experience also has taught us about the concept of machine learning, where our creations "learn" by storing pieces of information about the decision-making process that it went through and had not previously been stored resulted in a positive outcome, so that the additional information should lead to better decision making in the future. We must always keep in mind that real time information from sensors is critical to any good decision-making process, so there must be constant communication from the various sensors to the IProSys and this constant flow of information from superlative sensors with extremely quick responsive actions based on processing by the IProSys is what gives us the amazing spectacle of massive schools of fish or flocks of birds in very close proximity that move as if they have one central mind, performing maneuvers that would be catastrophic if attempted by similar number of our machines.

When a squirrel stores nuts or a bird builds a nest, not only are they performing similar decision-making processes again and again but for each specie they are referencing a similar store of information, and by observation we can deduce that each specie has identical stores of information that they reference to be able to make almost identical decisions.

Our ability to install logic into our creations is a very new capability, and thus it may be awhile before our abilities match the sophistication required to truly build fully autonomous robots, to say nothing of autonomous systems consisting of robots that are not directly related but with some interdependence akin to robots at every level of a supply chain from procuring raw materials to the manufacture of complete cars, drones or other sophisticated robots, of course none of which include distinctively unique characters and personalities that serve to make living animals unique.

With our inventions, we build all the code that is embedded in our robots and although we may like to think that our code is perfect, we are occasionally reminded that there can still be edge cases that reveal that our code contains bugs or may simply have encountered conditions for which we have not performed adequate testing. Our programming ability we must remember is less than one century old, in a few thousand years maybe, our cumulative skill and experience along with the tools that we employ may have advanced to the point that we have a lot more confidence in giving our creations "free will", based on the safeguards that we build not only into each robot, but also in the environment in which they operate, so that the system is not comprised of the robots but there are in effect guardrails implemented in the environment in which they operate. This is what we can observe in predator/prey as well as in the relationship between grazing animals and the areas in which they forage. This may also be in effect in the very long-term cycles that occurs with weather patterns and even sea levels on our planet.

Before intelligence, sensors

It is important to identify the components of any system in order to fully understand the inner workings of any complex unit or organism, it is analogous to understanding our machines, for example the engineering and full operation of an automobile regardless of the era in which it was created can only be fully understood if we fully delve into the operation of each of its systems, we must investigate its' electrical and electronics, its' suspension, internal combustion or system of motors for EVs as well as the various components that comprise its' frame, seats and even its' safety systems, if the automobile is of our latest designs we would also have to investigate its' computer systems; a full reverse engineering to have comprehensive knowledge of that machine. We would not accept that some unknown force just built this machine as the result of random happenstance or accept without question the capabilities and even the motivation of that unknown benevolent force, this however is what we accept when we vacillate between Nature as a random force that somehow give us elegant and sophisticated designs which our best and brightest strive to emulate.

We now know that one of the prerequisites to implementing any intelligent system is the design and installation of sensors. Sensors provide the means to collect the data which is then transmitted to the various processors that comprise the "brain" of the intelligent system. While this description may sound very mechanical, it is fundamental to any system that has to act autonomously, although we may not be able identify the various sensors which are essential to the efficient function of plants or animals, we definitely can recognize their critical impact on all organisms.

When we observe the functioning of these sensors throughout our ecosystem, it readily becomes apparent that unless these organisms have extremely advanced and sophisticated ability to self-design, there must be some external agent involved in first designing the various sensors as well as the systems to coordinate and process the myriads of data constantly being reviewed and collected.

Collection of data requires memory in which to collect and retain information, there must also be "programs" that not only initiate the collection of data into memory but also will direct the transmission of the collected data to the processing centers. So, sensors themselves must have memory and dedicated "programs" to decipher relevant information that has to be captured and processed locally or forwarded to more complex decision-making centers, which may or may not assimilate data from other sensors e.g. sight, sound, smell etc.

We also know that there must be pathways for the collected data to be transmitted to the various processing centers, currently most if not all of the instrumentation required are unknown to us for living organisms, we currently collectively attribute the design of these highly sophisticated sensors which are specifically designed for the needs of differing species, the mechanisms used to collect data well as the methods of transmission and processing to an amorphous and all-knowing and benevolent "Nature".

Let's take a dive into some of the more common sensors that exist in plants and animals that are fundamental to the proliferation of II which we refer to as instinctive behaviors.

There are sensors which may be implemented differently but perform the same functions in plants and animals, common function sensors would not be in our sense binary as in on vs off but be able to distinguish many different states between extremes, some of the essential would enable the ability to sense up versus down, humidity, temperature, light, chemical, gravitational, pressure and others which we may not yet be aware, while the functionality may be the same, it is obvious that specific implementations varies immensely, especially between plants and animals.

Our present experience with robotics reveals the type of sophisticated infrastructure and programming required for even elementary automation, and therefore the scope of work required to both build the infrastructure (in living cells, memory requirements etc.) to implement the equivalents of some of our more common robotic sensors such as gyroscopes, pressure, light, sound, chemical, tilt and those that equate to GPS but still evade our research and continue to amaze us in the independence from any known or discernible data sources.

Sensors are highly sophisticated dedicated pieces of equipment in any machine, in living organisms, they are highly specialized cells with often not fully understood mechanisms to both act on their own and transmit a range of input to more centralized processors. At the current time, we do not have knowledge on what quality or quantity of which assets are required for either sensors or the ancillary elements that constitute the framework that supports intelligence in any living organism, but we understand that one critical component is memory, memory which must hold instructions on a range of possible options based on input stimuli, again we currently have no method of quantifying or even identifying the memory that is utilized and we know even less about the method of storage and the information that comprise the instructions to the relevant sensor. We can surmise however that there is a process that enables the instructions to various sensors to be updated over time (machine learning) and that somehow this can be passed down to subsequent generations as evident in evolutionary cycles.

The very development of sensors indicates an in-depth knowledge of the stimuli to which the sensor must react or capture input, here again we see the "hand" of some intelligent force, which has intricate knowledge of that which will impact the organism, this intelligence is also displayed in the placement of sensors, additionally the "wiring up" of the sensors to local and central computing system. So in order for an organism to respond to sound or light, the designer of the sensor dedicated to collection of light or sound must have an understanding of the properties of light or sound, because as we know these are forms of energy with for example varied frequencies, so light sensors must be designed to respond to certain sector of the light spectrum, similarly with sound, an understanding of the sound spectrum is mandatory in order to build the sophisticated sensors we see in various plants and animals.

As of this writing we cannot with absolute certainty affix a number to the types of sensors that make up the full sensory capability of all plants and animals, we can observe within our environment plants and animals responding to seemingly invisible or currently undetectable stimuli as in the cases of animals before natural disasters, where none of our current instrumentation can identify the source of that which causes plants and animals to respond in predictable ways.

Until we invest in the research required, we may not be able to determine all the sensors that have been developed and which infers that the designer of these sensors, had advanced knowledge of the forces and other forms of stimuli that would have an effect on each type of sensor. There may also be complex sensor activity where the response by one type of sensor may have a domino effect on others, resulting in complex series of unconscious activities that benefit the host organism.

Sensors also play a prominent role in the generation of mutations. While there must be an ongoing internal communication involved in the generation of new cells between the "data processing" centers and the processes involved in the actual creation of new cells including the replication of existing DNA, based on input from the various sensors, it may be determined that changes in various genes or combination of genes may be needed to respond to persistent unexpected or "abnormal" signals being received by one or a group of sensors, this may result in the generation of several mutations over time with the successful mutants becoming the dominant strain over time with or without the elimination over time of less successful strains. It is noteworthy that each organisms' data warehouse may have ranges of what is "normal" for each type of sensor and this information may vary for each variant or mutation, so that as mutations occur, the intensity of the signal received from various sensors that are accepted as "normal" would change and as a result no processing to create additional mutations in response to "abnormal" signals would be required.

We must always keep in mind that real time information from sensors is critical to any good decision making process, but since the memory of a parent is not passed down to its offspring, the IProSys must also evaluate what information if any, should be used to update the II of the animal which would result in the upgrading of the instincts of the species from that that generation onwards, this in itself is in fact a form of mutation, so it may be possible that the experience of one organism can result in changes to the specie, however it may be that these changes are not "hardened" in that they are stored in some state where it is not referenced for future modification of physical or II until it "rises" in priority because multiple generations are hypothetically "making the same change request" since II modification would normally be accompanied by some changes in DNA and possibly in the specie specific II.

We may not know the full range of sensors that have been implemented in every form of life, but as they are discovered, we will become more enlightened as to the sophistication and incredible knowledge required to design these sensors, as well as gain insights into the knowledge of the forces that impact us that enabled the development of these sensors. As we gain insight in the abilities of plants and animals, both in their internal systems and the ways in which they interact with the ecosystem around them, there will be opportunities for us to gain insight into the various sensors that are part of their physical makeup. In time, with the advancement of our knowledge of these sensors, we will also recognize the mechanisms that are utilized for animals to achieve such feats as navigation as well as advance knowledge of impending weather or other unusual natural occurring events such as earthquakes etc.

These yet to be discovered sensors may include one or a group that is responsible for us having the sense of being watched or even others that bring on a sense of foreboding or elation in relation to events in the future. We must not lose sight however of the fact that in order to design a sensor, the designer of the sensor must have an intricate knowledge of whatever is to trigger the sensor being designed, in fact "it" may also have to be versed in the design of the actual IProSys which ultimately is responsible for processing the signals from the sensor, "it" would indeed be beyond genius!

One of the questions we should attempt to address is: do these sensors have interference immunity, and unlike turning off the receptors as may be possible with pain and pain killers, it may be that these sensors have been designed such that they cannot be "fooled", this is important in that the signals passed to the IProSys is always accurate as any interference of these sensors could pose an existential threat to entire species, imagine if the location sensors were somehow impaired so that the entire species go off in the wrong direction!

The issue of interference immunity is important because of the immensely detrimental consequences to affected individual and even entire species, but it also because demonstrates the advanced intelligence of the designer in the ability to not just account for primary purpose of the sensor but also to build such a robust architecture that it is immune from interference when designed and into the future. This also indirectly indicates the quality of "code" that drives the IProSys now and into the future.

One of the issues that we may be dealing with for quite some time, is that our instruments may be relatively so crude when compared to the forces that activate the various sensors in living organisms as well as the various transmission between the sensors, the IProSys and the other parts of the system like muscles or organs may be on such a scale that our instruments are currently incapable of detecting any of these forces, it is as if we are trying to look at nano particles with our naked eyes.

Just as we have many kinds of information being transmitted from our devices, like radio signals, cell phone messages etc. our world may be awash in signals that we currently cannot detect but for which all living organisms have sensors, thus it may be in time that future civilizations will look back at us as the dawn of recognizing this information ocean that we all exist in but that we were not aware of and thus could not make use of the resources around us, this is analogous to our cell phone signals around uncontacted tribes in the Amazon. It is likely we will be able to detect and monitor the operations of the IProSys only when our instruments approach the sensitivity of animals like the echidna, platypus or certain sharks that can sense that electrical energy that is the result of activity of the IProSys.

In time as our instruments become more sensitive, we may realize that there are additional sensors that are common to many animals, which we may also share but because the signals that are received are of such low intensity that we may have become accustomed to ignoring them or they may be easily drowned out by more intense signals from other sensors, it may be that with practice we may be able to develop our abilities even without sensitive instruments to hone in and focus on the signals that are presented by these sensors.

Just as there is a variation in the sensitivity between individuals of the recognized senses of sight, hearing, smell, taste and touch, there may be those individuals who have a much more heightened awareness of the signals received by these not yet recognized sensors. Even the ability to have a sense of being watched which many animals appear to react to, may be the result of one of these yet undiscovered sensors, it is only when our instruments have gotten sensitive enough to detect the real time information that guides "animal GPS" will we be able to definitely exclude the possibility of information being present around us, but being indiscernible to us.

It should be one of our high priority objectives to create instruments that can detect some of the signals that we know exist but currently cannot detect, because we can recognize that these signals happen at a scale that is currently too low for any of our instruments, when however we achieve the ability to detect these signals, our civilization would benefit from literally having instruments that can "tune in" on nature at a level that only living organisms can, this would lead to a greater understanding of why certain events that may currently appear mysterious occur, much more sensitive instruments will also enable us to listen in or observe the minute forces that occur within all living organisms and influence life itself.

Debunking the "Nature did it" myth

Our science advances due to the dedicated work of logical individuals, inferences are based on what has been proven before and any projections or assumptions are normally related to this established foundation, there are however notable exceptions. In the area of intelligence however it appears that our best and brightest have a serious deficit in associating what we know with what we observe. In our current state of advancement, we know that information has to be stored in memory, we also know that continual, recurrent logical decisions can only be the result of "good" information that is accessible to the facilitate the decision making process however, we have now firmly established an seemingly all powerful, all knowing entity that has no form and no direct relationship with us, yet seems to engineer systems that are for our benefit, while firmly denouncing the existence of any such entity!

One clue that indicates any concept or item is unserious is the inability of those who reference it to consistently and accurately define it, this is exactly the case with "Nature", its' state apparently varies as it is alternatively referenced as a phenomenon, a thing or even a place. Then there are always references to "Mother Nature" which, actually seems to reference the unknown creator and may be the most accurate based on what we now know about intelligence, as the programmer of intelligence must also be the creator of DNA or work very closely with, suffice it to say the specifications for the memory into which the permanent logic that comprises instinct must be known since each DNA configuration (specie) gets a uniquely engineered (yes, with quite some reusability) logic set embedded in its' permanent memory.

The acknowledgement that Nature is the architect of everything within our ecosystem is tacit acceptance that there is irrefutable evidence of intelligent design. Discussions of evolutionary theory has till now, focused on the development or advancement of physical traits, what has been missing but is like the silent elephant in the room is recognition that everything observed by Sir Charles Darwin as well as our observations of nature indicate that there are logical progressions everywhere in nature, what we have failed to address is that logical actions must be the product of intelligence as distinct from random events which demonstrably results in chaos. There is also no evidence of any relics that can be definitely attributed to random events in the evolutionary record, past or present.

Even more prominent in our consciousness than the assumption that the appearance of geometrical patterns is a definite indicator of intelligence, we must recognize that without intelligence all actions especially of living organisms will be chaotic, so we must recognize that from the unicellular organisms, every action is the product of one or more decisions and this decision making happens even within every organism, even when actions are controlled by chemical reactions, there must be decisions of the composition and timing of the production of said chemicals.

All of our sciences are based on either empirical data or commonly accepted principles based on readily observable, repeatable patterns for which we do not currently have the means to produce data that can be independently verified. Notable exceptions however are psychiatry and psychology, it is unfortunate however to hear almost all scientist acquiesce the source of the logic that dominates everything in our ecosystem to a rather mysterious, benevolent force called Nature which designed and created all things (not just the "living" by our standards), is responsible for the current state and actions within our ecosystem, is responsible for the creating and implanting of all II which we observe as "instincts" in all living thing.

It is a more pragmatic view that Nature is not an object in itself but more likely the result of the actions and processes designed by that the same source that created and embedded II in all living things. We may be taking some humanistic license when we presume that any creator would have an interest in its' creations…

There are literally millions of examples of the sophisticated logic that went into both the design of plants and animals and in the symbiosis between organisms that we must accept that it is just plain too easy and lazy to ignore all the intelligent indications and attribute the wonders of our ecosystem to happenstance. We currently accept that even the occurrence of straight lines and angular shapes in any regular pattern points to the work of intelligent beings whether there is current evidence of their presence. When we observe the Puffer Fish nest or the nest of almost any bird, we would not assert that over time, random forces could create such and intricate design, now that we recognize the sophistication of intelligence, it is inconceivable that we would continue to assert or accept that random forces could design and implant intelligence that is tailor made and close coupled to specific DNA configurations in every specie.

It is therefore ironic that even our best and brightest have without objection adopted the notion that all the intelligence we observe in Nature is somehow the product of or resulting from a series of random actions.

Just as we recognize that our development and application of intelligence is the defining capability that transforms our human operated machines into autonomous or semi-autonomous robots, we must recognize that intelligence is what guides any living thing to perform in predictable patterns.

Some of the activities with which are familiar definitely do not fit into any category of natural progression from one normal stage of development to another, examples metamorphosis, hibernation, aestivation and many others, these are outliers that seem more designed to demonstrate the incredible powers of a creator.

Many of these processes seem to break any rules that anyone would try to establish based on looking at any one sector or flora or fauna. If it was natural for eggs to develop into young animals inside the body, then it would be unnatural eggs to develop into young outside the body, if it was natural for eggs to be fertilized inside the body, then it would be unnatural to have fish fertilization in the sea, lakes and rivers, if it was natural for eggs to develop into young animals, it is unnatural to have metamorphosis in butterflies or frogs, if it was natural for animals to have their brains above their feet or hearts, then it is unnatural that we have bats, if it is natural for plants or animals to die if they did not ingest food for weeks, it is unnatural the ability of reptiles to survive for months without food, if it was natural for reptiles, then it is unnatural for mammals to hibernate, the ability of some marsupials to suspend the development of embryos are more examples of almost extreme unnatural abilities. There are literally millions of examples of unnatural processes that we can observe everywhere in Nature! These examples which are way more extreme that for example us building machines that could alternatively soar in the atmosphere and directly visit the depths of the oceans.

What we can observe is that the forces that forged the species across the spectrum of our ecosystem sometimes seemingly broke any "rules" that could be established by viewing any slice of the range of species as if to demonstrate that with infinite knowledge and the ability to manipulate both the DNA code and embed appropriate intrinsic intelligence, anything could be natural.

The very recognition that intelligence was required for the very design, placement and functioning of organs as well as the autonomic systems that are part of all living things, coupled with the knowledge that specie specific II is embedded in all living things, now that we understand what it takes to embed intelligence and transform something from an immobile form to an active autonomous machine which cannot approach the intelligence and complexity of any living organism, we must discard all notions that Nature is a random unintelligent force.

The definitive evidence that has been likely unconsciously ignored was due to the fact that intelligence was not considered as an attribute that should be investigated, whenever we discovered a new specie, we may have been amazed at the physical attributes, what can be categorized as the tangible assets, with no attention whatsoever being paid to the intangible assets; the intelligence, and especially the II or instincts of the organism, any review of the II would have led us to recognize that not only had "Nature" endowed it with special characteristics but "Nature" also gave it logical intelligence to make best use of its' special attributes. This examination would have quickly discredited any notion of random, as the specialized logic we observe everywhere absolutely could not the product of random processes and we recognize that none of these animals taught themselves and somehow imbedded their instincts so that it is passed down to all offspring in their specie.

THE SIMPLE AND THE COMPLEX

There is one additional phenomenon that even as a standalone issue argues against the notion that evolution is independent of some logical force, it is the degradation of the ability of animals to sustain themselves independent of their parents as we move up the evolutionary path. There are two aspects to this counter intuitive result, what we can observe is that as animals become more complex or what may be referred to as advanced or developed, instead of retaining and even becoming more independent at birth, we see the more advanced species having the least physically developed young at birth so that, this is coupled with seemingly a diminished amount of survival instincts, to the extent that the young of the most "advanced" species not only require adult care, nurturing and tutelage for years before they could survive on their own, also notable is the fact that there is less intrinsic intelligence evident in their normal activities, humans would not know what to eat or even to provide themselves with shelter comparable to spiders, birds or other "lower" animals.

What is readily apparent is that animals lower on the evolutionary scale are more equipped to survive and thrive by a combination of much more advanced physical development at "birth" as well as having a more comprehensive version of II included. So even though they may have a much shorter lifespan, flies, spiders, lizards etc. are much more capable of finding food sources and in many cases have the II required to be excellent hunters when their species depend on consuming other living organisms. Whereas a case can be made relative to the rate of development, it is highly improbable that the II required to thrive independently would be diminished as the animal become more developed in evolutionary terms so much so that the higher primates require substantial periods of learning from others to be able to survive and thrive.

What we can observe is that as we go up the evolutionary tree, the autonomous systems appear to be the equivalent, autonomous functions like digestions, immunity and other life support functions all appear undiminished, what appears to be is that the ability to learn new skills from observing others as well as active tutelage is heightened but the amount of detailed instinctive knowledge is diminished or it may be that the logical processes of any animal with free will can supersede any instinctive impulses or reactions, this may be the state that we exist in, where our teachings based on custom or logic becomes so dominant that II is almost completely suppressed.

We may indeed have some of the advanced sensors that we can recognize are involved in the decision-making processes of the "lower" animals but, in the "higher" animals, conscious actions may be dictated mostly if not completely by knowledge learned, if this is so, in the future we may learn to be what is currently referred to as being more intuitive by learning to "listen" to the more subtle signals that are received by all our sensors.

It does make some sense that more complex organisms like those "higher" up on the evolutionary tree be more in need of nurturing if they are to be expected to themselves be nurturers, but this does not hold true in many instances, but the apparent loss of the complex and sophisticated II that we can observe from the simple to the complex, argues against a natural progression of physical capabilities.

The tight coupling of physical attributes and specifically modified II that seems to raise the ability of the simple to survive and thrive based purely on their embedded intelligence, where the more complex also requires nurturing or they would simply perish, argues against a purely natural, or random development process.

There are of course, many examples of "hybrid" cases, where the association of many "simple" animals can live cohesively in very complex structures, with separation of duties which indicate complex planning, we see this in the many animals, notably insect that live in large colonies. This separation of duties which of course requires specific versions of II to be embedded in different animals within the same colony so that they perform distinctively different and specialized roles, these may be accompanied by changes in the DNA of these groups in order to make them most efficient in these roles, as in the various defenders, caretakers etc. in insect colonies.

Even rules that we may want to establish for the operation of the simple and complex animals as we see do not aways apply, as both simple and complex animals may live in large communities where they may operate independently or with some form of collaborative instinct. There is also beyond that II that is embedded in each individual, another form that lends itself to members of communities being able to "teach" others, this we can observe in bees for example who can transfer information on food sources, so effectively simple animals can have II that instructs them to "make a note" of the location of food and then to "tell the others" back at the hive, significantly blurring the lines between the simple and the complex.

Intrinsic Intelligence (II)/Instinct: Programming with Autonomy

Intrinsic Intelligence (II) is best defined as "**evident knowledge that is not obtained from external sources**", we observe II manifested every day in plants and animals as those behaviors which we call instincts. This is not any behavior that is learned by any means, not by any direct instructions or from observing others, so the learned activities of higher primates and others are not to be considered in this category of intelligence. II is however present in all life forms, it depends on sophisticated sensors that capture relevant stimuli which is then passed to local or centralized processing centers, this then results in decisions to engage other sensors and or specialized cells to perform some action if necessary. From unicellular organisms like the amoebae to the activities of stem cells or those we observe everywhere in plants and animals in Nature, the process is the same and consists of the following steps: **detection, processing and reaction**.

While we are currently unable to determine what constitutes the memory in living organisms, we know that it is a critical component, not only is it needed for the actual processing of information received from the sensors but, it is also for storing the vast quantities of information (the data warehouses) that must be interrogated to make the best decision based on information collected from all available sensors. We can safely theorize that as organisms become more complex, they have a larger range of sensors and thus have much more memory available both for processing "live" information from each sensor or sensor group as well as for storing information relative to each sensor/sensor group and possibly for cross checking and verification using all forms of and/or logic to effectively make the best decision based on all available received information, so, every organism needs adequate memory to store lots of information and uses additional memory to process information received from its' sensors to make the best decision and initiate the "instinctive" response.

Although this sounds rather simplistic, in order for the significance of this statement to not be understated, it is worth repeating that "*every intelligent action occurs as a result of logic*"

Some examples may be necessary here to highlight the execution of the detection, process, react steps that are always part of instinctive actions which are based on II:

I.   Amoebae: even these unicellular organism may have a range of sensors but for this example, pressure and chemical sensors may be the types required, the pressure and/or chemical sensors would detect the presence of other items or organisms, this would result in processing the information to determine whether the object or organism encountered could be an item of food, the chemical receptors would also transmit enough information so that a determination of whether the encounter is with another amoeba, and unless this species is cannibalistic, detecting its' own type would not trigger an attempt to engulf and consume the detected organism, also, based on the information obtained in the detection phase, the processing can also conclude that evasive action is the best course to be initiated as a reaction, either way, we can observe the result of the processing phase by the logical action that follows.

II.   Plants: plants have a variety of sensors but if we take as example a "typical" tree and a grass, both would have pressure, light, moisture, chemical and other (temperature/gravitational/?) sensors. In this very complex and highly intelligent processing system, based on input from its' various sensors, the plant proceeds to germinate, the direction of roots as well as the initial leaves are the result of processing signals from several sets of sensors, that trees "decide" to send down roots to considerable depths and grass to only produce fibrous roots, both are the product of information stored in their data warehouses that are interrogated as the various sensors that are involved in the growth of the plants continues to send signals for processing during the entire life of the plant. Exactly which sensors are preeminent during the various phases of plant growth we may only know after resources are dedicated to research. We know that the physical characteristics are determined by data encoded in the DNA of the plant but all the dynamic aspects of the plant; direction of growth of leaves and roots, when to grow quickly or slowly, when to produce flowers that are used to attract pollinators, all these decisions are made at the behest of the various sensors. So based on our observation of plants, we must recognize that there are sensors for events that we currently do not fully understand such as, plants that hibernate and the various sensors that cause the triggering of changes in their roots, leaves and internal systems.

III. The development of a fertilized egg is an excellent example of advanced intelligence directing the process that starts of as a series of cell divisions but soon is transformed into producing some of the most sophisticated creatures and all their supporting systems. The timing of the change from duplication to specialization and creation of these amazing structures and the related support systems obviously are not done by DNA and cannot be the product of any random process.

We must consider that since the IProSys mechanism has to be integral to unicellular organisms as well as embryonic cells that it may in fact be a part of every cell similar to DNA, so that it would form part of a networked architecture that along with developed nerves and the brain in appropriately configured animals, contribute to the sensory intelligence of every organism. An intracellular IProSys would account for even plants being able to quickly respond in defense or target resources to recuperative activities when damaged.

The activities that we attribute to instinct obviously are not related to the internal functions of living things but more to those actions that we can observe, and our deductions so far have not accounted for the obvious logical actions as being the product of a distinct attribute, just as any physical characteristic that we can view.

It is illogical to deduce that all the rational processing we observe in animal instinct is the product of random evolution, especially when there is abundant evidence that there are instinctive actions that correlate with lunar activities, being cognizant of the fact that animals do not consciously pass information of their experiences down to their offspring.

We must also consider that instinct can be so strong that it can override what we would consider the natural reaction based on the circumstances, this is evident in the action of deer fawn when they stay immobile in the face of loud moving equipment that occasionally run over them. There are even cooperative instincts as we observe between multiple members of a specie, one prominent example is that displayed by mouth brooders, several fish breeds the male may stop eating while it protects eggs and the young fish in its mouth, the cooperative instinct works in that both the parent and all the young know how to react whenever danger is present, the young all rush into the mouth of the parent who accepts them and keeps them until the perceived danger has passed, definitely not a learned activity.

DNA is responsible for the physical characteristics of us all but, autonomous Intrinsic Intelligence specifically designed and implemented in every life form is not only responsible for the evolution every specie but also comprises the instinctive knowledge that guide their daily actions as well as the autonomic life sustaining activities of all living organisms.

Examinations of Intrinsic Intelligence (II)

These are just some of the literally millions of examples of II which we refer to as instincts that we can observe almost daily around us or at least we can view in some form of media. They all demonstrate decision making that is NOT learned from tutelage or observing others, and so these are excellent demonstrations of the intelligence that is somehow embedded in the memories of that subjects.

Because these actions now seem natural, should not lessen the fact that these actions are the product of well- designed and highly detailed "programming" which may use a variety of sensors to produce activities that are very predictable and well documented, each analogous to an exceedingly, well designed and meticulously tested software implementation.

Every example described below could be illustrated in a rather simplified way in a program flowchart, of course none would be complete as we may not know all the various "inputs" that are referenced in the various decision points, but we can be assured that without a decision-making process, none of the activities would be so consistent and repeatable.

We must keep in mind that without intelligence, none of these processes make all the right decisions at the right time, and for intelligent processing, there must be sensors that send the appropriate signals, often time based to enable the starting and stopping of very intricate processes such as the generation of certain types of cells, chemicals etc. for signaling other sensors or to the IProSys for coordinating disparate processes within plants or animals.

It is also very important to keep in mind that unlike us, these plants and animals do not have a written record from which to build from one generation to the next and in most cases, there is very little if any parental contact for the young animals referenced to learn by observation, also, the knowledge that is apparent from the behaviors we observe has a very high consistency across entire species, so it must be specie specific II.

Procreation:

We can observe that most animals that actively seek mates during activities that lead to the creation of the next generation of their species go through a period of physical and behavioral transformations that are triggered by the production of various chemicals within them. Very much like the cell division that occurs with a fertilized egg, this process only occurs after animals reach a certain age and those transformations that often occur only during certain periods which may be influenced by external factors. All the internal changes that occur within members of every specie has to be directed by some intelligence, the synchronization of the physical readiness to breed to certain periods whether based on seasons or the availability of necessary resources all point to intelligent decisions and the product of the interaction of multiple sensory inputs with a highly developed and sophisticated IProSys.

What we can recognize is that there may be multiple type of sensors, some internal and some external that contribute to the IProSys that it should initiate the chemical changes, again there is nothing chaotic or random, what we always observe is a predictable sequence of actions, which must be directed by cogent intelligence.

Even the recognition that the aggression that many species display to ensure that the strongest have the best chance to be the ones that pass on most of the genes to the next generation is in itself the product of logic and intelligence. The question is who or what determined that the "code" that results in the competition for breeding rights be included as opposed to in the case of random or chaotic behavior, that members of species would simply breed with any other member of the specie to bring forth the next generation.

Imprinting:

This is a behavior that we observe in some of the "higher" or "more evolved" animals, it may be the result of a newborn animal using visual sensors to identify another animal that is in its' vicinity soon after "birth" and consciously storing that image, then immediately creating a bond of recognition, trust and dependence on whatever that it first laid its' eyes upon, this can of course lead to young animals creating this "bond" with others including those that are of other species, this can also occur with the maternal animal somehow identifying certain traits of the newborn animal and storing it so that it can uniquely identify its' young from among hundreds or even thousands of others that may seem identical.

The obvious "programming" that is apparent with imprinting obviously relies on either the young or parent animal to respond to specific sensors then store a certain amount of transient data into their "permanent" of some type of memory that is retained for some period, during which a comparison is made "does the input received by certain sensor(s) match what stored information" "IF" yes "THEN" perform actions that are pertinent to parent/child relationship "ELSE" ignore and move on to comparison with next potential "relative" to repeat the IF/THEN/ELSE process.

Stem cells:

From their discovery around 1981, we have learned that there is a class of cells that can manufacture many other types of cells, and if this was a random process, we would indeed have chaos wherever stem cells are found but what we have discovered in fact is rather than being agents of chaos, is that stem cells are extremely useful in the regeneration of various types of cells under controlled conditions. The very fact that creating a specific environment can lead to an expected outcome in living organisms indicates intelligent behavior.

As we have stated elsewhere in this text, the IProSys must exist in every cell and in the case of a fertilized egg, it directs the timing and manufacture of specific cells and systems towards the end product of an entirely new animal of the specie, what we can observe in stem cells is the intelligence inherent in the IProSys can recognize through information received form relevant sensors that what is required is not the generation of an entire organism but only certain types of cells, so here again we can recognize intelligent processing of information with resulting actions or directing of targeted actions that demonstrate the amazing capabilities and exceptional intelligence of the IProSys.

Immune Systems:

The immune system provides an excellent demonstration of the power of the IProSys, the living organism demonstrates no conscious control over a system that operates for its benefit, the identification of invasive or foreign we must remember requires comparisons with a known database whether of chemicals or other organisms, so there must be sensors that pass signals to the IProSys which is on a real-time basis is constantly checking "**IF** sensor X signal is normal **THEN** nop (No Operation) **ELSE** signal appropriate routine based on signal to perform additional checks/initiate activity", this we observe as the production of various chemicals or the dispatching of certain cells that react very much like bees or ants in defense of their queen and/or nest, the IProSys may also by its increased activity be partly responsible for the temperature fluctuation of affected organism that we detect as being feverish, what is significant though is that even though all the agents may have be activated, their activities ceases the moment the IProSys is shut down, which we recognize as the death of the organism.

Digestive Systems:

Similar to the operation of the immune system, organisms do not consciously control their digestive systems, even reptiles that may have significant changes in their rate of digestion, this is all controlled by the IProSys, based on signals from appropriate sensors, we can observe the production of various enzymes to aid in the breakdown of foods, the absorption, distribution and even the production of waste products are all under the control and direction of the IProSys, so much so that even with all the enzymes etc. being present if the IProSys is shut down, all these dependent operations, even of the seemingly independent enzymes ceases, demonstrating their reliance on the IProSys being in the "on" state.

Kangaroo suspended pregnancy:

The details of this remarkable process demonstrate a major deviation of from what has been established by the observation of other higher evolved animals, it is the type of process that demonstrates supernatural abilities by certain species, well, except they occur naturally in this specie. It involves the female getting pregnant but suspending the development of the embryo while she is taking care of an older sibling! This is an example of a truly amazing feat of self-control if any animal could consciously make that decision and enact all the physiological changes within their bodies to enact such a remarkable feat, were it a conscious process, would it not be something that our species would be able to emulate?

It is apparent that they have the appropriate sensors whether chemical or otherwise that by certain patterns of signal transmission to the IProSys, indicate that care of a young is ongoing, the processing that would result in whatever changes would have to be the result of very intelligent and sophisticated processing that we would be proud to emulate, to say nothing of actually knowing when the appropriate signals are received, to initiate and implement whatever processes are required to maintain the pregnancy in this state of suspended animation. Of course, sensors would also have to indicate to the IProSys when it is appropriate to initiate processes to terminate the suspension essentially resuming a normal pregnancy!

Kangaroo Joeys:

Kangaroo babies are born very much looking like embryos; however, they have a very high success rate navigating from the birth canal into the pouch where they continue their development after attaching themselves to their mothers' mammary gland and suckling for several months. Closer examination of what is required for this amazing feat when we consider the size and physical development of the fetus like kangaroo at birth, we can see that this entire process is powered and driven by II, we can see the sensors/IProSys feedback loop being engages such that based on input from various sensors, the IProSys directs the appropriate muscles to propel this seemingly helpless animal into its' mothers' pouch, once there the same sensory reaction loop propels it to attach to a nipple, the mothers' IProSys on detecting the attachment of the baby joey, directs the enlargement of the nipple so that the baby kangaroo remains attached so that it can continue its' development.

Wombat:

Most of the examples highlighted in this chapter require us to observe certain behavioral patterns and deduce or make inferences about the intelligence required to design and embed information into many of the lower animals, this example however, is about a physical design that is like all designs in nature logical, but this stands out as it breaks a pattern for a specific reason. The wombat is the only marsupial and so it for certain periods of the life of its' young, the joey is carried in a pouch, the wombat is markedly different from other marsupials in that it burrows into the ground and so if it had a regular forward-facing pouch like the other marsupials, it would be filled with dirt likely killing any occupants during its' natural activities and so the wombat has a backward facing pouch.

These "exceptions" or "edge cases" serve to highlight evident intelligence inherent not only with the II embedded in all creatures but to the logical design implicit in all fauna or flora, and since we accept that no creature can consciously modify its' own DNA to enable changes to be passed along to its' offspring, we must conclude that there is an intelligent agent which has full access to both DNA and II involved.

Puffer fish nest:

This is one example that has to be seen to be believed, fortunately we can see multiple examples if we simply search "puffer fish nest", and it immediately conveys the obvious intelligence that is required for a fish to create this amazing three-dimensional structure with its' precise geometrical pattern, on a lighter note: there are no young puffer fishes around taking notes. This is not something that this specie does every day so again we must recognize that there are a range of sensors that trigger the ability, and that the information to create this intricate pattern must be stored in the memory of every relevant member of the species. This aptly demonstrates that there must be embedding of information and processing of that information to enable this specie to demonstrate what we must now recognize.

It absolutely contradicts any random assertion because that would almost certainly entail some "great, great, great grandfather" suddenly having an epiphany and also being able to somehow store it in its' own IProSys so that it is passed down along with the DNA for all members of its specie, as this cannot be the case, we must accept that the puffer fish nest is a demonstration of the power of some advanced intelligence to develop patterns and then to implant the knowledge into what are considered animals of lower intelligence to demonstrate the amazing capabilities of whomever or whatever was able to embed such amazing II into this specie.

Positioning prey for swallowing (snakes, birds etc.):

There are many examples of us driving species to or almost to extinction by interrupting their normal feeding patterns, specifically if we start poisoning the prey of some animal with the result that the predator consumes the poisoned prey and itself dies, there is no evidence that others in the specie somehow get the message that they will also die if they continue to consume the poisoned prey, even if the behavior of the prey is different when poisoned. Recognizing this fact, we continue to observe that all animals that consume their prey whole, takes the precaution or ensures that they swallow their prey head first, this is not behavior that can be learned from observing the fallen comrades who inadvertently made the wrong choice but it the result of knowledge that is embedded in their memories, and obviously not by any relatives that got it wrong.

Coconut crab:

Another example of extraordinary intelligence at work can be seen in the actions of the coconut crab, an animal that in its adult phase cannot swim but females "know" that they must deposit their eggs into the beaches close to where they live. So, the female actually puts itself at risk of drowning in order to deposit its eggs in the water, and the sophisticated programming does not end with the II embedded in the female, there is also evident sensors that trigger all the eggs to hatch immediately upon coming in contact with water, even if the egg "shells" are designed to simply dissolve, these facts must be available to the designer to enable this entire operation to complete successfully.

Bird Nests:

Both the construction and location of bird nests are definite demonstrations of sophisticated II, even though there may be some imprinting of the location and some sense of the materials used, the actual process of nest building as well as the procuring of the materials demonstrate the embedded II that is specific to the specie. The varieties of nests that are built and the locations chosen by individual species makes it very clear that there is not a linear progression or even that the traits are not passed down from one specie to another, wheat we can observe is the demonstration of choices being made based on embedded II, which are all displays of intelligence.

Scent marking:

The development of specialized organs that enable the dispensing of substances that transmit information about animals as well as the specialized development of other organs that contain the sensors to decode the messages left by others is so complex and focused that when first encountered should definitively have indicated that there could be no random generation of any parts of these systems and should have led us to the only logical conclusion that these systems were intelligently designed. The manufacturing of specialized compounds, the composition that enables specific information to be dispensed, as well as the location of both the distributors and receptors for these compounds are still beyond our best and brightest, clear indicators that there is nothing random about anything in these systems.

The locations where scent markings occur as well as the frequency are also indications of a specialized intelligence that cannot be passed down consciously without an oral or written tradition. These systems are just beautifully designed examples that are almost like markers that intelligent observers should recognize as being elegantly designed and implemented systems, now that we are aware of the logic that has to be executed to implement both the physical and computations that has to occur to detect and interpret the information passed by chemical means, it is apparent the level of sophistication required to build the sensors and the advanced logic that has to be included in the IProSys for the successful implementation of what we observe every day in nature.

Hyacinth macaws:

These are beautiful birds that have fantastic adaptation to be able to make use of a food source that most other birds find impossible, the adaptations present are very pronounced and almost targeted, if we are to accept that this bird did not encounter this food source as a challenge and make a decision to modify itself to be able to overcome the challenge, then how can we explain the logical physiological changes between this bird, namely the muscle concentration, the dexterous tongue and feet that enables it to maneuver and crack the shells of the very hard palm nuts so that it can consume the energy rich contents.

These are the sort of specialization that we can observe every day in nature but, when we recognize that for a physical modification of either plant or animal, the actual DNA has to modified as well as there is often an "upgrade" to the instinct or knowledge that is required for one specie to make use of a natural resource, it becomes clear that there are internal decision-making processes that autonomously facilitate whatever changes are required. When there are modifications required that involves multiple species whether between different animal species or between plants and animals, it is inconceivable that any symbiotic process can be developed independent of shared knowledge. This will be addressed in another section.

Animals that feign injury:

This is an instinctive behavior that is observed in multiple animal species that is intended to "trick" a predator into following it away from other more vulnerable components of its' species such as eggs or offspring. Again, here we have a demonstration of highly developed intelligence demonstrated by the feigning of an injury with the explicit intention of protecting something more vulnerable, all of this totally unlearned but somehow embedded in the memory so that the actions required can be recalled and executed flawlessly. As nothing can repeat multiple times what if does not know, these animals are demonstrating awareness, and intent to protect something more vulnerable by performing from "memory" a series of actions. The only question is who or what implanted this knowledge so that when triggered by some sensors that there is danger present, they can recall and enact the actions we observe.

Deer Fawn:

We have observed that young fawns will lie still even when predators are nearby, they have been known to maintain their position even when farm machinery approaches and so quite a few fawns are killed when they are born and "hidden" in fields that are to be processed by farm machinery. We now also recognize that fawns have almost no scent, so that as long as they do not move, unless stumbled upon, they cannot be detected by predators, they however have all the faculties to detect that which may harm them but that II during that stage of development likely has logic that goes something like this "***IF parent detected THEN get up, greet, acknowledge etc. ELSE stay immobile***" and so this would not be a case of frozen by fear but because of the reflexive action based on signals from the IProSys is "stay immobile". Of course, this reaction can only be effective because the fawn does not give away its' location by having some detectable odor, so in this case, this is a logical decision, one that does not apply to many more animals, where the "ELSE" process would be very different.

Ocean creatures spawning coordination with lunar cycles:

This phenomenon which has been observed raises questions about the types of sensors with which these organisms are equipped, as the actions are obviously directed by some intelligent agent within the organism, that receives input from sensors that monitors something else which tracks moon phases or they themselves have sensors that track the phases of the moon, as there has to be a some "preparatory" work that has to be done prior to having mature eggs ready to be released with the correct tide, so whatever sensors are utilized, the information provided by them enables the IProSys to direct and coordinate the other changes at the appropriate times so that mature eggs are ready for the tide.

The fact that something whether internal to the sea creatures or external but in their environment has sensors to monitor moon phases opens a pandoras box of possibilities regarding yet undiscovered sensors, and since we know that in order to design a sensor one must have in-depth knowledge of that which has to be monitored, that is to say in order to design a light sensor one must understand the energy imparted when light falls on an object, and we may have to be more specific in that we may only be interested in our sensor reacting to the entire spectrum from infrared to ultraviolet or only certain frequencies, designing sensors that exist in aquatic environments where the turbidity varies and their role is to monitor celestial bodies is a challenge that can only be met by truly advanced intelligence.

Another option which has to be quickly dismissed is that these creatures can sense a change in the movement of other characteristic in the water that indicates tidal movements and so their IProSys can respond accordingly, this however may be possible by us due to our oral and written record keeping, because change can only be detected by comparison between to states, so these creatures will have to record multiple states of their environment and perform the relevant comparisons to determine the appropriate timing, not impossible but highly improbable.

Animal navigation:

This is an area of which we know so little that there is a lot of speculation as to how animals have accomplished what prior to us using Global Positioning Satellites was the domain of experienced and expert navigators. Obviously, the sensors that animals use have been developed and perfected with interference immunity so that the animals that depend on these sensors can do so with complete confidence.

There are many aspects that we now understand conceptually, which makes it even more amazing when we see them flawlessly executed by animals, one of these is setting a "home" location, we see this in many animals like sea turtles and salmon where the "home" location is established and stored for their entire lives, so that when some sensory inputs based on age and readiness to create the next generation are received by the IProSys, it then processes that read the "home" location and directs all systems to start the animal moving to this location for breeding, egg laying activities.

This ability to set "home" location is not only done in the very young animals, there are some examples like racing pigeons and bees where the home location can change in some cases very frequently as in the case of bees that are rented out to provide pollination services. Pollination services is a business that requires many bees to be brought to some farm location, possibly because there is a dearth of pollinators due to manmade or other causes, obviously the rented bees who have been transplanted with their queen have the amazing ability to reset their "home" location, because they cannot only depend on the scent of their queen's pheromones when upwind of her location.

As we know today based on our method of implementing location services, sensors must have real time data, so whatever system is implemented in animals we would assume also uses real time data, for corrections to be made based on current location, but whatever the details are of the implementation that applies to animals (and speculation about fish like the great whites etc. using the stars just seem too "magical" – be lost all day then resume at night? does each fish on its own decide or is there II based on the stars implanted?!) the data needed by the location sensors and the IProSys ability to make the comparisons based on information passed by the location sensor, lends itself to sophisticated intelligence that is unlikely the product of random processes.

It is notable that whatever these animal sensors use to navigate, whether based around the earth's magnetic field or otherwise, that signal is present at all times. Our current civilization if very advanced but, here again as our instruments become more sensitive in the coming years whether that be decades, centuries or longer, there will be a time when we will be able to detect and navigate using these ever-present signals, in that future, our current implementation of GPS using satellites will seem almost primitive, as of course that entire system will become obsolete.

Prey Detection:

Some of the very interesting and difficult to explain developments in the evolution of animals are the sensors that extend the abilities of animals to locate prey and are outside the "normal" sight, scent and sound. Although the evolution of the sensors that enable excellent sight and hearing were they to occur as the product of random happenstance would be amazing, the development by certain predators of sensors that can detect the heat signature or the small electrical currents (likely those that occur during the work of the IProSys) definitely seem to defy any notion of randomness when we recognize that in order to design a sensor, one must have intricate detailed knowledge of the energy that has to be sensed.

It would make sense that the designer of the IProSys, who would be intimately aware of the various levels of energy that occur as a result of the work being done by the IProSys, would be the ideal designer of sensors to detect the energy signature as the result of the work of the IProSys, even directional scent detection by the forked tongue design of snakes are all indicative of comprehensive intelligence as no snake needs training in the scent its prey and it does not get misled by the scent of animals on which it does not feed, similarly sharks and other animals that have sensors to detect the electrical signatures must also detect other animals but because their sensors all have interference immunity, they are able to target only prey animals.

Venom:

The development of venom in animals that have this capability bely the notion of randomness of evolution because along with the sophistication of the venom synthesized, is the related and also specialized "equipment" used to ensure efficient application of the venom, again no relic of chaotic randomness. The very development of venom is a demonstration of knowledge inherent in one specie of the vulnerability of another specie, which lends itself to an intelligence outside of either specie with intimate knowledge of both, as a snake for example could not know without some knowledge passed by its own testing group for example of the effect on different species so that it could be assured that its venom could be used in obtaining food.

Corolla Spider:

Here is one example of animal behavior where there is such specialized usage of II that it can stand alone and provide incontrovertible proof that II has to be imparted to an animal by some super intelligent source. The Corolla spider catches prey by digging and lying in wait in a hole, it is what it does prior to waiting in this hole, in fact turning a simple hiding spot in a trap that truly stands out and demonstrates a fascinating implementation of advanced II. In order to turn its hiding place into a wonderful trap, the Corolla spider places quartz crystals around the hole and attaches a web from the quartz crystal that extends into its hiding place, it stands on the attached we waiting for any possible prey animal to disturb the quartz crystal, vibrations from the quartz are transmitted along the web to the spider's feet, it then springs into action to catch its prey.

This is such a fascinating case that enables us to highlight several distinct aspects that all fly in the face of any "randomness" and points to external intelligence embedding II into an animal, we have the fact that the spider is able to identify quartz crystals, that a spider and it should not matter how many millions of years since any one spider does not live a million years or does it pass down the notes of its research to other members of the specie, as we recognize in order to identify, we must make comparisons to stored information, we must compare some characteristics to those we have stored, so the Corolla spider must have stored information on some unique characteristics of quartz and obviously that stored information must include the ability of quartz to vibrate in a manner that allows enough energy to be transmitted the distance from the quartz crystal outside its' hole to feet.

We know how to identify trees, buildings because we must be given information to be able to make some visual comparison, but any of our other senses can be useful in identification of objects around us, so in this case, we may not know what senses are being used by the spider to identify the quartz crystals but we know that it has to constantly be making comparisons with its stored information, what it "knows" to be quartz with what it encounters as it digs, even the size of the chosen crystals may have been compared to its internal database, all of which provides evidence of imparted II.

Even after the trap has been set, the quartz crystals have be laid out around the hole where it lies in wait, there is also the specialization of the sensors that detect the vibrations that are transmitted via the web, the sensor must also be able to distinguish whether the vibration is caused by the wind or some large animal which could be a possible predator, so there has to be very fine tuning of the sensors so that the IProSys can direct the appropriate action.

Clothing the young:

DNA we learn are learning changes during the lifetime of an organism, this epigenetic quality cannot be random else it would be the source of chaos in the individual plant or animal. What we can observe is that for many species, this ability works to bestow upon the young special coverings that are designed in concert with the quality of care meted out by the adult animals, includes both the coloring as well as the physical characteristics of the material that is used to "clothe" the young animal, the combination of the local environment, the care meted out by the parent and the "coat" with which the young animal is equipped is always complementary, which definitely indicates "design" considerations.

We can also observe instinctive caring by young by both parents when the need for warmth is critical and the environment poses an existential risk to the young if both parents were to be away for extended periods obtaining food, the fact that it is "programmed " into both parents that they cannot both leave at the same time, the fact that in those species, there is shared common knowledge, demonstrates that there must be a source for that information that embedded that within every member of that specie and not others.

So, in order for the young to have adequate warmth during their critical vulnerable stage, we see a few common but different strategies, "nature" providing the warmth where there is one parent caring for the young, or instilling the knowledge that leads to cooperation between both parents when extended exposure can be fatal to the young, logical decisions that can only be implanted in DNA and/or II by a benevolent third party.

Paternal mouthbrooders:

There are several species of fish that demonstrate this behavior which as with so many other of these examples, require complementary II be implemented in multiple members of a specie. In these instances, the male of the specie uses their mouths to keep eggs and/or young fish in their mouths whenever there is a risk to the safety of their young. This behavior may result in the parent not being able to eat for some period, "knowing" that eating would result in death to their offspring.

That the father knows to use its mouth as a safe haven and the young to know to enter the mouth of their protector, obviously requires complementary intelligence on both sides, how this could evolve out of random happenstance is something that most would agree is very highly improbable if not impossible, and yet this display of coordinated intelligence is normal in nature.

Net casting spider:

As their name implies, these spiders construct a net from their web and throws it over their victims before attacking and consuming it, even the quality of the silk used to create the net is different and specialized for it function, here we have both the implementation of special modification to the genetic makeup of the specie so that it is able to create a special quality web along with the sophistication to create the net, and the specialized knowledge of how to "cast" it over prey, this spider may also "mark" the spot where its net will be cast naturally with white fecal dropping so that it creates a target and thus knows when to cast its' net as a potential victim crosses the target.

The examples above all provide circumstantial but overwhelming proof that some extremely advanced intelligence was involved in the complementary modification of the DNA and implanting specialized II into the permanent memories of each of the species mentioned. It is definitely illogical to assume the modification of the DNA and implantation of memory that enables specific instinctual behavior is coincidental.

Extraordinary specializations

We take for granted physical forms and events that are logical or seem to make sense, because we are an intelligent specie however, one of the tenets of our scientific literature is that our ecosystem is the product of random happenstance, if however we delve into just a few of the numerous examples of cohesive, cogent systems that exist within every specie and the innumerable interspecies collaborations, it readily becomes apparent that not only is there a logical design of species but it appears that there is somewhere a collective knowledge of every specie so that design decisions appears to have been made where not only is collaboration beneficial but in many cases they are essential to the well-being and even the existence of disparate species.

There are examples of phenomena in nature that simply defies any rules that we may try to form simply by observing regular patterns, what we see are outliers that seem to fly in the face of any normal rules that we could form based on normal progression as would be evident in "normal" evolution without the twists and turns and breaking with convention that actually lends itself to some creator force breaking with "simple" convention, it is as if to demonstrate its' mastery by creating edge cases that break the rules and yet conform to natural patterns of species existing and thriving.

It is as if we had someone with such expert ability that they chose to not just do normal things with expert flair but also to demonstrate their supreme talents by "showing off" and purposefully breaking all the known rules but still have amazingly successful results, such are the cases we can observe with animals that hibernate or aestivate as well as the process of metamorphosis and others that we will touch upon in the following examples. It must therefore be apparent that the intrinsic intelligence evident within the animals that have these abilities or have these processes as part of their lifecycles must have additional sensors and processing to enable the successful execution of these extraordinary processes.

Hibernation/Aestivation:

For animals equipped with the special programs that enable hibernation or aestivation, there obviously may specialized physiological changes that have to accompany the special hibernation/aestivation internal programming. There must be a range of sensors that indicate the conditions that require the animal to enter these states of suspended animation and special conservation of resources etc. are approaching, the combination of the signals being received by relevant sensors then causes the IProSys to send the signals to relevant organs to start whatever activities that precede this state of suspended animation.

It is important to note that these actions by the IProSys are so pronounced that it changes the behavior of the animal, as with procreation, the IProSys may trigger the production of certain hormones etc. that causes the animals to modify their behavior. Here again we see the sensor, IProSys, reaction loop that results in decisions that are not consciously directed by the animal but in which their actions are completely "instinctive".

Metamorphosis:

This is one especially interesting example of the complex IProSys that must exist within the single cell from which a butterfly, frog or any animal that goes through this process. The DNA which acts as the blueprint of the physical characteristics of the animal will contain the information that dictates the animals' physical characteristics throughout its' lifecycle, but since DNA is not time based or has the ability to process itself, it is the domain of the IProSys to monitor the cell division/cell creation from the single cell to the first stage of the animals lifecycle using the relevant information in the DNA, so the intelligence as to what the caterpillar looks like is in the database but what part of the DNA/database is to be used to control the physical characteristics of the caterpillar is made by the IProSys, the processing of signals from sensors based on the cell development that leads to the initiation and control of processes that lead to the formation of the pupae is controlled by signals processed by the IProSys, which also reads the DNA to determine and direct the activities that result in the specific characteristics of the pupae, the IProSys also is responsible for monitoring, initiating and directing the formation of the final butterfly or moth.

What is almost beyond comprehension is that the IProSys exists in the very first cell and although it may be a "bootstrap" version that leads to more advanced versions as cell division and specialization occurs, it is more than likely that the IProSys is a complete system, capable of reading DNA and directing the creation of an entire plant or animal even when the adult animal has completely different physical characteristics to any of its' previous forms in its' lifecycle.

Coloration:

Plant and animal coloration bring beauty and wonder to our world, there is much about which we must wonder indeed. Coloration in plants serve many purposes, some we can ascertain and others we cannot fully, but it is obvious when plants use colors to attract animals that there is indeed a third party designer as not only does the DNA of the plant need be modified, but changes have to be introduced into the various animals so that they have an affinity to the colors of the plants with which they have an affiliation, and although in some animals, the plants that provide food and which depends on these specific animals for pollination or seed dispersal are not normally taught by parents, so this information has to be stored in that memory which constitutes the instinct of the animal.

Animals like plants may use colors for many purposes some of which are: attracting mates, to establish or emphasize their superiority over others of the same species, for camouflage, for assist with internal temperature modulation or to make others be aware that they are to be avoided, this last option of others being aware to stay away from certain colored animals, is the one option that most definitely demonstrates a sort of corporative instinct because there is no animal symposium that agrees on which colors all animals should be wary of, or, the colors that certain animals should "wear" so that others will know that they should stay away from animals with certain colors.

In all cases of instinctive behavior, there is evidence that there are internal databases which are used for comparisons, because as discussed elsewhere in this book, we only "know" what a house or a car looks like because we are familiar with images of them, for animals to know that a poison frog is an animal to avoid, or aquatic animals to know to heed the warning form the Blue-Ringed octopus, there must a process of comparison between and image in their "instinctive memory" and the animal in their presence that results in "Avoid" signals being triggered because there is no information sharing as literally, none makes the mistake and lives to tell the tale!

The occurrence and behavior of color changing (chromatophores) as well as texture changing cells (papillae) in some animals like chameleons and octopuses, demonstrate some of the capabilities of the distributed, networked IProSys, where local sensors signals can be interpreted with the response being directed to ensure local color and or texture matching that is not dependent on what is in front of the eyes of the animal as it is obvious that the eyes even if they could perceive the depth of color and texture is limited in their scope of view. These amazing abilities being attribute to random and being limited to certain species only do indeed seem oxymoronic, even if the ability to encode these abilities were dependent on the individual animals, it would be highly likely that many other species would have the same initiative to include these abilities in their arsenal of capabilities, so instead we all are told that Nature found it fit to bestow these capabilities to only these selected species, which indeed may be closer to the truth.

The ability of some animals to change color seasonally or to match changes in their environment are all very logical actions, the ability to encode that capability in their DNA is a highly technical skill that most would agree is beyond the capability of any of these animals, even the growing and shedding of the coats are all very logical processes based on the changes in their environments and these changes seem normal to us because we are logical creatures but, this is another glaring example of us superimposing our logical thinking on what we have been told is based on random happenstance, and any investigation of the "intangible" process that lead to what we observe, leads to intelligence in the designing of these processes, in order for this to be occurring across a wide variety of animals, simultaneously, it must be an imbedded autonomous, sensor driven process.

Animal mimicry:

Another example of the flaunting of supreme mastery that is not subject to any conformity can be observed in the species of animals that have taken to mimicry of others, some even mimic multiple animals both in sound and appearance. This all demonstrates very sophisticated modifications to both DNA and II that enables these fantastic results to be manifested.

From birds that mimic others and even our voices, to the mimic octopus that can mimic multiple animals, animals that use parts of their anatomy to attract prey like the alligator snapping turtle or the South African puff adder or even the Spicebush swallowtail caterpillar or hawk moth caterpillar which can imitate bird droppings or mimic the appearance of snakes to prevent predation, what we can observe is the perfect implementation of DNA modification which enables sound or look necessary as well as the embedding of II so that the animal "knows" how to use the "assets" which the DNA modification made possible.

Human fingerprints:

This is a very notable characteristic of humans and bears mentioning because fingerprints have been identified as a reliable method to differentiate individuals, as our sciences have advanced, we now recognize that even identical twins who were formed from a single egg and thus have exactly the same DNA and yet have different fingerprints.

In one of our most advanced technologies; software engineering, we use very specific methods to ensure that certain unique keys that are to be used globally, such as the Media Access Control address (MAC ids) assigned to a Network Interface Controller (NIC) are not duplicated anywhere so that we can be sure that each machine connected to the internet can be uniquely identified, we do this by a convention which is universally accepted. If however, each company was to randomly choose a number for each device they manufactured, we would surely have the chaos that is associated with random, with human fingerprints however, since it is accepted that no two individuals have the same fingerprint, it is appropriate to question what system is used to ensure there is no duplication of fingerprints in the entire history of mankind, including when two individuals are created from the same egg!

Even if we assume a system akin to our system of assigning each company a unique number and within each company every device created gets a unique number and the MAC id is a combination of the unique company number and the company specific unique number, this would not work with identical twins, obviously having the same parents and being created from the same egg, so undoubtedly whatever system used is more sophisticated than even our methods for creating another unique identifier the Globally/Universally Unique Identifier (GUID/UUID) which has no guarantee of being unique forever.

So, if we really accept that fingerprints are never duplicated, logically, we must also accept that somewhere there is a "beautiful" algorithm that ensures that these human "hashes" are never duplicated.

Symbiosis:

If we accept that plants and animals do not have the ability to self-modify at will, then we must accept that all instances of symbiosis must be engineered by some "third party". We must remember that logic is normal for us but in the context of random mutations, it should be a totally coincidental event that should also not persist. At this point we should be familiar with the documentary "<u>In the Mind of Plants</u>" which like this book recognizes that plants like animals have memory and perform actions based on logic. The following are just a few of the plethora of examples of symbiotic relationships and in each we can witness the intelligence that went into both the physical modifications as well as some insight into the II that plays a part in the actions of all parties in these mutually beneficial relationships.

There are many symbiotic plant to plant relationships, where it may appear that it may be totally coincidental, but close examination has shown that both benefit and thrive due to the association, examples include the mesquite and cacti in the Sonoran Desert, one plant providing needed shade, the other providing a source of nutrition in the form of its fallen leaves which decompose and provide nutrition in a harsh landscape, there are also mutually beneficial relationships with organisms that are known as nitrogen-fixing bacteria that benefit from food provided by the plants as a result of photosynthesis while the bacterial activities produces nitrates that the plants need to thrive.

Some animal/plant and animal/animal symbiotic relationship definitely stretch any insinuation that the relationship was born out of random coincidence. Examples of plant/animal symbiotic relationships are difficult to reconcile with as is the case of the acacia and ant relationship where the tree DNA and II had to be modified so that the tips of its' leaves would secrete nectar which feed ants who defend those leaves and so effectively the parts of the plant needed for photosynthesis, so whereas it seems normal that plants create nectar in flowers to attract pollinators, we have this continuing theme of exceptions to conformity that demonstrates true mastery of creation.

It may be argued that almost all pollinators have a symbiotic relationship with plants as many plants need cross pollination to successfully produce fruit and so just pollination by wind would not suffice to produce fruit, and the dispersal of fruit by animals including us, are yet another example of symbiosis, as the plants cannot put their fruits on planes, trains and automobiles to distribute the seeds.

As with plant/plant and animal/plant symbiotic relationships there are many animal/animal mutually beneficial collaborations that demonstrate that there must have been some synchronized modification of DNA and II across multiple species for the relationship to be as deeply intertwined as we can observe. One example is the herding of aphids by ants, one would expect that normally ants would kill and eat aphids but we can observe that the aphids are essentially "herded" and protected by the ants and they provide the ants with a food source that they produce and excrete for their protectors, we can also see another example that demonstrate the extreme modification as well as the flaunting of conformity in the relationship between the various clownfish and anemones, where protection is afforded to the clownfish whereas other species would be killed by the poisonous barbs from the anemone in return for providing cleaning services as well as their providing sustenance via their droppings as well as any small fish that may be attracted to the "resident" clownfish.

Clownfish DNA has to be modified so that they produce a layer of mucus, again we must remember that the wrong fish, that is small fish without the protective mucus would be killed and digested, no going home to start a project of modifying one's own DNA for the next generation(s), so we once again can recognize the hand of benevolent third party resulting not only in the DNA of the clownfish being modified, but its II is also updated so the it "knows" that it can reside within close proximity of the anemones with no negative consequences.

Of course, there is a plethora of symbiotic relationships out in nature but in all cases, we can observe intelligent design at work, often it is also apparent the specific collaborative DNA and II modifications that had to be implemented to make the relationship special and mutually beneficial to both species.

Diving Bell Spider:

Here again is another example of random evolution gone awry or a creator demonstrating its mastery of design and breaking any simple rules that could have been established by our observation of most other spiders, this is accomplished by modifying the DNA to enable it to have waterproof hairs along with significant modifications to the "basic" II that is embedded within these arachnids to have it "choose" to eschew the normal habitat of all its closest relatives and make spend almost its entire life under water, getting its oxygen from a captured bubble of air that it creates, relying on its oxygen store to be refreshed through a natural process of diffusion due to the relative concentrations of oxygen in the bubble and the surrounding water. There must be corresponding modifications to the sensors that enable the spider to detect and capture its food.

Sargassum Fish:

These are fish that live among floating mats of Sargassum algae, their truly amazing feature is that they are covered with fleshy weed-like appendages and filaments to resemble algal fronds and branches, even their colors mimic the coloration of the algae in which they live. It is also mind blowing that we accept the explanation that they have over time decided to adopt the look of the seaweed where they live, this makes as much sense as "in time monkeys will adopt the look of the trees in which they live!", or "we spend so much time in cars that eventually we will look like cars!", it also begs the question of "did the Sargassum algae play any part" in the decision of the fish, to modify its' DNA to mimic the seaweed? What is much more feasible but not conceivable until now that we realize that there must be a source for the intelligence that governs the behavior of the fish in this, and every environment, is that the source of intelligence must also have access to the DNA in all cases as there is always a tight coupling between the form and function in nature.

Bombardier beetles:

These beetles are wonderful examples of edge cases that break the normal expectations for insects and seem to be another case of an animal incorporating characteristics that show off the mastery of its creator by showing that the seemingly impossible or at least that which defies conventional logic can be accomplished and made to seem normal, of course, for us mere mortals, it evokes many questions but, it should again be an example of something that is definitely not the product of any random process, there is just too much that could have gone wrong in the construction of these physical specimens!

These beetles have been endowed with the ability to manufacture explosives! They are also physically equipped with a blast chamber to house the initiation of their generated explosion! So, their IProSys has the instructions to direct the production of chemicals that when combined creates an explosion that generates as much heat as boiling water, this it ejects towards whatever danger it perceived.

Without seeing this design in nature, most would agree that it is inconceivable, and if this design was to be presented to most project managers, it would be rejected as not feasible and yet, all the strategic modifications have been made to the DNA as well as the supporting installation of II has been successfully implemented in this amazing "product".

Electric Eels:

Like the Bombardier Beetle, here too appears what can be construed as another flaunting of supreme mastery that is not subject to any conformity, how else can we view the adroit modification of DNA and the embedding of associated II that enables an aquatic animal to use or harness the power of electricity? The very development of the special electrocyctes that enable the generation of electricity, the processing that occurs as well as eels seeming "knowledge" that leaping out of the water has a more devastating effect than when the energy is discharged solely in the water, all point to instinct that is tailored to be most effective within those species that have this ability. As with all things in nature, here we have a very tight coupling of specialized physical form with an expertly designed II system that together functions flawlessly.

Carnivorous plants:

These plants are all rule breakers, and it is difficult to comprehend what these plants accomplish without attributing a truly advanced intelligence to these individual plants or to their designer, the variety of elaborate and detailed strategies that are employed by stationary plants to capture and consume mobile animals involve such deception and trickery that it is very clear that a lot of planning went into each design, how we were led to believe that this level of sophistication could have occurred randomly is almost as bid a job of deception and trickery!

We can observe in these species' multiple methods of capturing animals, from traps that cause the target animal to fall into a digestive liquid, to snap traps, fly paper, false exits or other methods to wear out the animal and keep it trapped so that it can be digested by the plant. Again, we see that flaunting of mastery without being constrained by conformity: plants that trap and "eats" animals, digestion that occurs on the outside of an organism!

What we don't see but what is evident is the decision making that that actively has to absorb the nutrients from whatever is captured as well as in some instances the resetting of the trap mechanism to capture other victims, so there is an active process of responding to the various sensors as well as performing all the housekeeping functions to effectively capture its nutrition.

Northern Fulmar:

The one aspect of this sea bird that ensures that it is of relevance to this list of special modifications to both DNA and II that we can observe, is this birds' ability to projectile vomit a stinking oil that damages the feathers of other birds that may seek to prey on its eggs or young. As with all forms of chemical defense in animals, there has to be not just the special modifications that are required to generate the chemicals, but there has to be a parallel development or modification of the II so that the animal has the awareness of the potent defense capability, part of the modification of its instinct is also the knowledge of how to deploy and even the range for which the deterrent is effective, so that animals that deploy these defenses also have to be updated as to when relative to distance of their intended target, here again we see that specialized II implementation that indicates implementation by some truly advanced intelligence.

Animal mineral acquisition:

There is this phenomenon that has be observed with multiple species of animals, these animals may travel significant distances and the result is that there is quite an accumulation of animals at certain locations where they all appear to be imbibing dirt that is rich in certain minerals, this is not done to quench thirst or to sate hunger but the fact that multiple animals travel to these locations must lead to questions along the lines of "what forces are at play here?", (since we know why, the animals benefit from the intake of minerals), "what sensors indicated the presence of the minerals?", "what sensors indicated that obtaining these minerals from dirt as opposed to being part of their regular diet was beneficial?", "is there something programmed into their instinct that guided them (or at least the first of them) to this mineral rich source?", these are questions that we may resolve in time as these actions are definitely driven by some store of information and directed by some inane logic which we currently do not understand but, provide more evidence of some common II which in its' sophistication includes information that can result in finding unique and unexpected sources of supporting life.

Mangroves:

These plants have some amazing adaptations for live in an environment that is deadly for so many other plants, each adaptation demonstrates a profound understanding of the challenge faced and the solution to the challenge displays a brilliantly devised solution that we must remember results in changes manifested in the DNA of the plant as well as II to coordinate real time responses throughout the living plant. These include filtering salt as it enters their roots as the salinity of the water where they grow would kill most plants, and accounting for whatever extra salt bypasses their root filters either by storing it in old bark or excreting it from their old leaves, so that as these are shed, they effectively remove excess salt.

Because they live in muddy low oxygen soil, they have special roots and absorb needed oxygen in various ways with roots that are designed to account for the changes in sea level between tides, they even breathe through special openings that close during high tides, as we have found out they are great for that area between land and sea and are wonderful protectors of our shores from storms and tidal surges, they even build up the land mass, their seeds are buoyant and can travel for extended periods before quickly establishing new groves and protecting new lands.

Now all these modifications to the DNA and II are logical, because if we had the ability, it is what we would do to create a plant with such versatility, if we could modify the DNA and implant II that directs all the logical processes that control the biological processes that make this wonderful plant a success.

Bioluminescence:

The process of animals creating bioluminescence has been extensively researched and documented and this serves as our normal procedure to examine and explain the tangible side of what we observe, what we have not till now sought to decipher is how what we observe is tied to the behaviors of animals that exhibit bioluminescence. When we do as with all other special modifications, we will observe that there is a correlation between the bioluminescence that we observe and the behavior of the animal we will notice that there is a logical change in behavior that occurs every member of the specie, this demonstrate that there is a change in the II that is "loaded" into every member of the specie.

Up till now we have not been recognizing the change in behavior of a specie when there is a change in the anatomy or other physiological change because we accepted without question that specialized logic was inherent in all creatures, but even a cursory view will bring the realization that customized logic cannot be the product of random occurrence and so we now have to review much if not all of our previous work in which the logic and intelligence components were previously ignored.

IS THE IPROSYS THE SPARK OF LIFE?

As we have seen throughout this text, the IProSys is the intelligence that is integral to every cell and is responsible for all the logical decisions that direct everything from cell division to reactive initiation of other processes/activities based on input from all the various sensors, it is logical to assume that without all the prerequisites in place to support that activities of the IProSys that cellular activities may become chaotic or even cease to function, in effect the cell become aberrant or dies.

The IProSys within each cell forms part of what is analogous to our networked computer system with both local and centralized processing of certain signals from the sensors that are distributed throughout the organism, because of its quasi networked design, there is awareness of the state of the individual that is transmitted and maintained, so that reactions to adverse events in plants and animals can be registered with coordinated responses and updated information as to the state of the various sensors and the overall health of the individual being "known" throughout the "network".

All the autonomic processing that occurs within organisms are directed by the IProSys, so the functioning of our immune system, digestion etc. are all controlled by and subject to information stored in the IProSys, since it is also very much involved in all II activities, regenerative actions that occur whilst asleep etc. are all under the control of this networked intelligence system.

Somewhat analogous to our computer processors, we are able to recognize that computers are doing work by external indications such as heat, or the results of their work being displayed to us, but the challenge if a computer processor dies in a multiprocessor system, we have to identify and replace that processor even if the others in the array can detect and not "bypass" the defective processor, but, our computer systems are orders of magnitude behind the capabilities of the IProSys with its' latent abilities to process signals from disparate sophisticated sensors.

We know that there is a correlation between the IProSys and electricity which has led to the invention of effective devices like the electrocardiogram (EKG/ECG) and the defibrillator but, by comparison with the sophistication and low energy levels of the IProSys, these machines may be incredibly crude, large, cumbersome and inefficient by comparison, it may be a while before we are able to design instruments that can detect and measure or monitor the activities of the IProSys.

Our current inability to detect or interact intelligently with the IProSys is likely due to the fact that we do not currently have instruments that can measure or monitor events at the scale that they occur, it is like trying to observe nano particles with our naked eyes, but we can always observe the consistent results and relatively accurately deduce the underlying activities.

It may be a while before we are able to not only observe the relatively stationary physical components that make up living organisms, the cells, chemicals and the blueprint that dictates the physical aspects but also the more dynamic and even esoteric components such as memory, the centers of processing and even the methods of signaling by sensors, the local and coordinated processing by the IProSys, and one day even be able to access the content of memory or even to interact with the IProSys, we may then truly understand the majesty of the who or what designed and implemented these extremely sophisticated systems with such mastery that they all work autonomously with unmatched complexities masked by apparent simplicity.

It does appear that there are requirements for the proper functioning of the IProSys, we recognize these as the requirements for proper cell functioning, the systems that transport nutrients to the cells are vital to the proper functioning of the IProSys.

The IProSys is required to direct and manage based on the sensor feedback loop all autonomic systems, although current literature of the discoveries we have made of these systems would indicate that autonomic processes are basically controlled by the presence of chemical agents, there is still that gap that does not explain the triggering of these chemicals, the controlling of timing and most importantly, the processes that are responsible for an immune response or digestion all terminate even though all the chemicals are in place once the IProSys has shut down, which we know as the death of the organism.

We may in time, be able to understand if, why and how the IProSys may be influenced by external factors that may include some things that are currently beyond logical explanations, these may include what are now considered to be mystical forces, ancient customs or exercises, we may even learn of non-invasive, non-chemical means to positively influence the operation of this critical system.

We may also in time definitively debunk or recognize as valid many of the ancient assertions about our relationship with our environment, our upcoming journey of self-discovery may be as revealing and exciting as anything we may encounter elsewhere in our ecosystem. Even the relationship between the brain and the rest of the distributed and networked IProSys may hold amazing discover, we may well be able to discern if the brain is really the seat of individual consciousness!

It is also significant that we now recognize that effectively the IProSys can operate independent of the brain, which of course it does in a fetus prior to the formation of the brain, but as a result of our medical advancement, we can observe this ability when due to illness or injury, we are able to maintain bodies with proper functioning of all other organs, effectively the spark of life throughout the body even when there is no sign of brain activity, demonstrating that the IProSys operates independently of "consciousness".

One of the possible "bread crumbs" is our discovery that all living things have DNA, again this leads to logical conclusions as using a common building block, would enable the building of a common IProSys that in a sense is "familiar" with DNA and so that there may even be common "modules" with "reusability" for many of the functions of life, especially if the IProSys is actually the spark that when extinguished, the state of the organism changes from alive to dead.

THE CREATOR ISSUE

The fundamental issue we have with asserting a definite creator to the order and intelligence we discover is one of accountability. There is this issue of accountability because it opens a very thorny question of "What is or should be our relationship with this Creator?", the inherent assumption being that, a creator would have some of our characteristics, and like us; have a desire to oversee its' creations.

As of now, we do not have the capability to build a self-regulating and self-sustaining ecosystem and as such, we have a need to monitor and control our "creations", as such, our expectation is that a creator would be establishing a similar type of "parental" relationship with us, this would be a major conflict with our sense of self determination based on our intelligence and our ability to establish control of many of that which surround us.

Projecting what we know about ourselves into a creator, would leave us with some accountability to whomever or whatever is that original Creator, this is the choice that most religions make, our scientific bodies on the other hand appear to have taken a hybrid choice; by recognizing that there is simply too much evidence of intelligence in the structure and design of everything within our ecosystem, along with the fact that there is no evidence of chaotic design and no relics of randomness anywhere in nature, all scientific fields now choose to refer to Nature or Mother Nature to account for the prevalent evidence of intelligence in design and operation of everything we observe in our ecosystem.

There are those who believe and experiment with recreating the environment that theoretically caused the transition from chemistry to biology thus bringing forth life, because we have not addressed intelligence as an inherent attribute of all successful life forms, they seem completely oblivious of the fact that even if we could simulate the creation of DNA from the basic elements, there is the even larger issue that we would somehow need to address; how would we introduce the intelligence that would enable the resultant "life" to transition from random chaotic actions to some semblance of intelligent behavior. At this point, this secondary issue does not have to be addressed, as first step one must be accomplished and the fact that there is a related step two apparently has not been recognized!

Now we can recognize that "life has to have logic", so no experiment in a lab would simulate life unless it is able to embed the logic so that the new lifeform can act totally independent of its' creator as well as be able to replicate itself with a complete copy of the logic to act with "free will" and be able to independently replicate itself with automatic embedding of said logic in every "copy" of itself.

Some things that we should all be aware of is that the intelligent agent or source that is responsible for the creation of DNA and the innumerable specialized II that exist throughout our ecosystem definitely demonstrates a mastery beyond anything that we could conceive, but it also should be apparent that there may be quite a bit of reusability of the "code" that is the IProSys, we should recognize this due to the universality of DNA across all lifeforms with our ecosystem, this indicates that the memory that is used to store the II that is also inherent in all living organisms.

One of the likely issues with a Creator could stem from the fact that we have not been able to envisage the development of a truly autonomous ecosystem and although we have become more familiar with many aspects of the components that go into making a self-regulating ecosystem, as the scope of the design and implementation comes into view, with all the varieties and complex interdependencies, it appears highly improbable that one source could engineer the complex balances of life and food chains that we observe in both the aquatic and terrestrial environments.

With the revelations of our sciences however, we can recognize the design of critical elements, and the sophisticated engineering techniques that are at work. The use of at least two common elements as reusable building blocks, one that dictates the physical configuration and one that contains logic that can be packaged together would be excellent starting points for building an intelligent system. Of course, it would help greatly to be the designer of both or have access to the full "specifications" so that each could be modified to create organisms that would seem impossible until they are discovered as occurs occasionally, but enable them to function successfully with all the variations that have been documented here and elsewhere. Not a simple task but given the keys and evolutionary times, we might be able to accomplish quite a bit.

So, it is up to the reader to contemplate and try to resolve whether the intelligence we observe is the product of random happenstance, and the really tough issue to resolve with this option is; can random create the comprehensive intelligence in the design of everything we observe around us, remembering of course that random equates to chaotic results, or could there be some pattern wherein random quietens and become regular patterns of logic that somehow over time can even be customized and implanted as specie specific instincts? Alternatively, there are the options for some intelligent agent or agents with obviously advanced capabilities eons more advanced than us that could "engineer" the comprehensive, interconnected ecosystem that we all call home.

THE PROMISE OF THE FUTURE

We will find in time that true sophistication to approach the levels in nature is only achievable by miniaturization, both with our ability to build sensors as well as being able to detect existing stimuli that occur in far too minute quantities for our current instruments, we know that the sensors in the forked tongue of snake transfer a sizeable amount of data to the IProSys from prey that may be a significant distance away, this accounts for snakes entering the houses of humans based on the presence of mice and rats, when we are able to create instruments of similar sensitivity, similarly there will be huge changes in our medical treatments and procedures based on our improved understanding of the IProSys and the mechanisms involved both in normal and diseases, so that the utility of the IProSys can be directly influenced.

The progress of our sciences in large part has taken on a straightforward and logical path; we observe, we seek to explain what we observe often with hypotheses, then we investigate. Investigations require appropriate tools and methods, if for nothing else but to ensure that verification can be performed independently by unaffiliated individuals or groups, this part of the process however cannot be performed until the appropriate tools are available.

We are very much aware that much of what occurs in nature and is apparent to other species happens at a level that we currently cannot adequately detect or measure, and although this flies in the face of us being more developed, it would appear that our sensors have been considerably dulled in comparison with other species with which we coexist.

All our senses are diminished considerable but, we have the ability to create sensors that capture and amplify signals that would affect the sensors of all the other animals, we simply need to invest the time and resources to be able to create sensors that simulate or even surpass those of the animals we observe, this would likely open up a whole new world of signals to which we had been previously oblivious, and many of the mysteries of the unknown will then begin to reveal themselves to our rational world.

When we have sensors that can match the capabilities of the most animals, with the senses we recognize, and especially those that give us insight into the IProSys we may enter the realm of detecting the signals that seem so mysterious as when animals seem to detect upcoming natural disasters, or seem able to detect when others are under severe stresses or even when animals seem able to detect undiagnosed ailments.

We will find that even our current most advanced medical therapies would look primitive when compared with an ability to monitor abnormalities and tailor responses to the actual IProSys. Being able to closely monitor that state and operation of the local IProSys will provide a window into our health that will open new horizons on what is possible with healthcare.

The correlation between the IProSys and the health and well-being on an individual when fully researched will take our medical sciences to unprecedented heights, as we may be able to make diagnoses earlier and when we are able to review any data directly related to the IProSys, our remedial process will become much more targeted, even the process of reviewing the results of remedial actions will become more focus and detailed, enabling even more precise interventions and better results in all medicals disciplines.

There are many headlines currently and there will be many more even to the point that serious examinations will have to take place due to the massive power consumption related to our use of AI, this of course is related to the significant increase in computing power required to make all the additional decisions and the additional computations based on the increased amount of data that has to be processed, so at some point there will be serious cost benefit considerations, however when we look at the massively superior ability of the IProSys, which of course has mastered miniaturization in both size and electrical energy across its supremely networked architecture, while processing even larger amounts of data from multiple sensors simultaneously, we may be able to realize significant saving with the inherent benefits to us, as well as the environment.

Disruptions of established conventions take time and even in the face of evidence, many organizations become very bureaucratic and insist on maintaining whatever is the status quo, fortunately we live in a time where no one group owns the right to investigate or stand in the way of those who would, with our current ability to collaborate and the almost instantaneous distribution of information, we may be able address many of the challenges that we will undoubtedly encounter on this road to another level of scientific discoveries.

Recognizing intelligence and the fact that up till now, we have had an almost total decoupling of logic from all the investigations and discoveries will result in many aspects of our life sciences having to be re-addressed, but that is a normal part of scientific advancement, we often have to re-evaluate, take a step back then move forward with a more complete, comprehensive and fundamentally secure basis for our sciences.

While it is unlikely that the question of the who or what is the definitive originator of both DNA and the "code" that sits at the base of the tree of life, it will become clearer as we examine all the evidence presented that the hypothesis of everything being the result of random happenstance will be discarded but without physical proof of that creator but with overwhelming circumstantial evidence, the case will certainly be very easily understood and believed but there will always be those who will debate the nature off, but will have to accede that there is indeed a creative force responsible due to the overwhelming circumstantial evidence presented by our examination of the logic and intelligence.

The greatest achievements of the future will be our decoding the secrets of the engineering of the past.